IMAGES OF THE UNIVERSE

BILDER AUS DEM UNIVERSUM

IMAGES DE L'UNIVERS

IMMAGINI DELL'UNIVERSO

IMÁGENES DEL UNIVERSO

Compiled & Edited by Govert Schilling

More titles in preparation. In addition to the Agile Rabbit series of book +CD-ROM sets, The Pepin Press publishes a wide range of books on art, design, architecture, applied art, and popular culture.

Please visit www.pepinpress.com for more information.

COLOPHON

CONTENTS

The Pepin Press | Agile Rabbit Editions
P.O. Box 10349
1001 EH Amsterdam, The Netherlands

Tel +31 20 420 20 21
Fax +31 20 420 11 52
mail@pepinpress.com
www.pepinpress.com

Compilation & text by Govert Schilling

Concept & design by Pepin van Roojen

Translations: LocTeam, Barcelona (Spanish, Italian
and French) and Sebastian Viebahn (German)
Copy-editing: Ros Horton (English)

ISBN
90 5768 067 X English/International
90 5768 087 4 Deutsch
90 5768 088 2 Français

10 9 8 7 6 5 4 3 2 1
2010 09 08 07 06 05

Manufactured in Singapore

INHALTSVERZEICHNIS
SOMMAIRE
INDICE
CONTENIDO

Free CD-Rom in the inside back cover

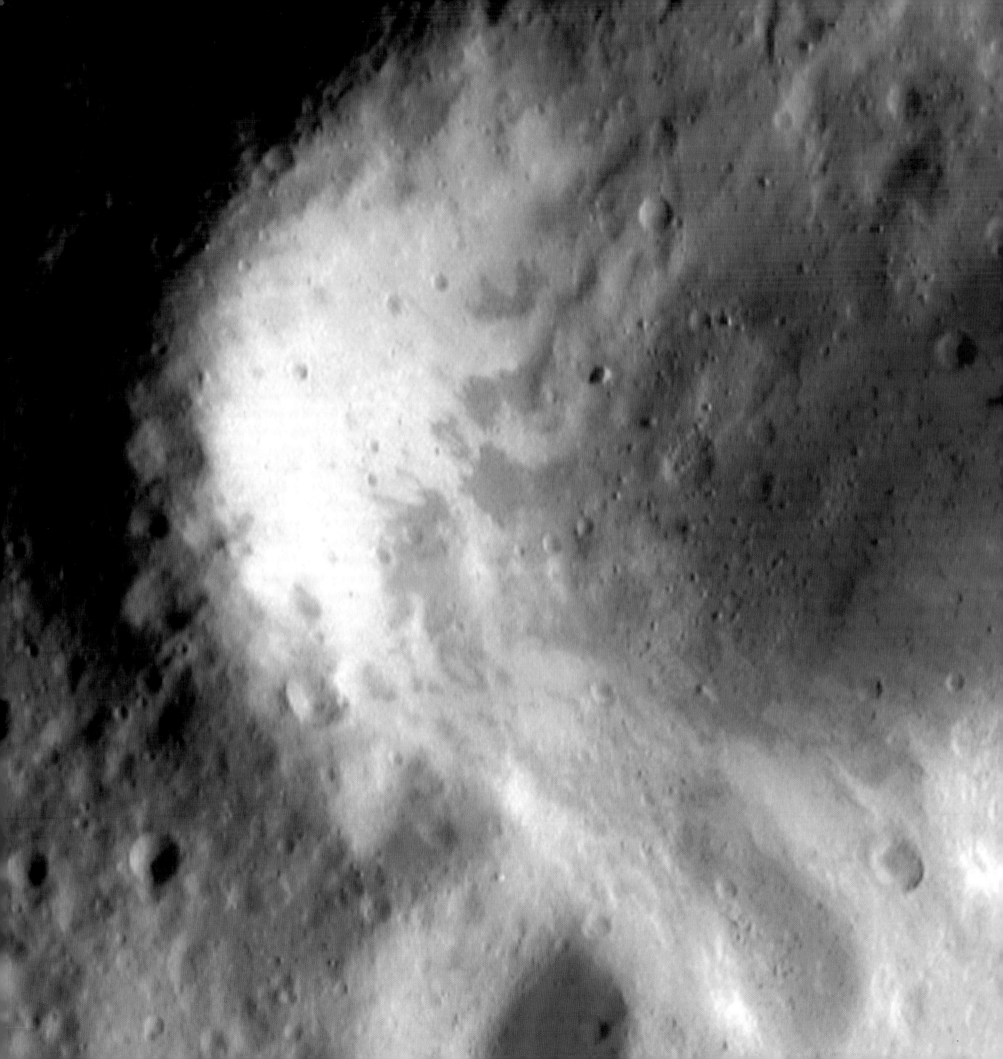

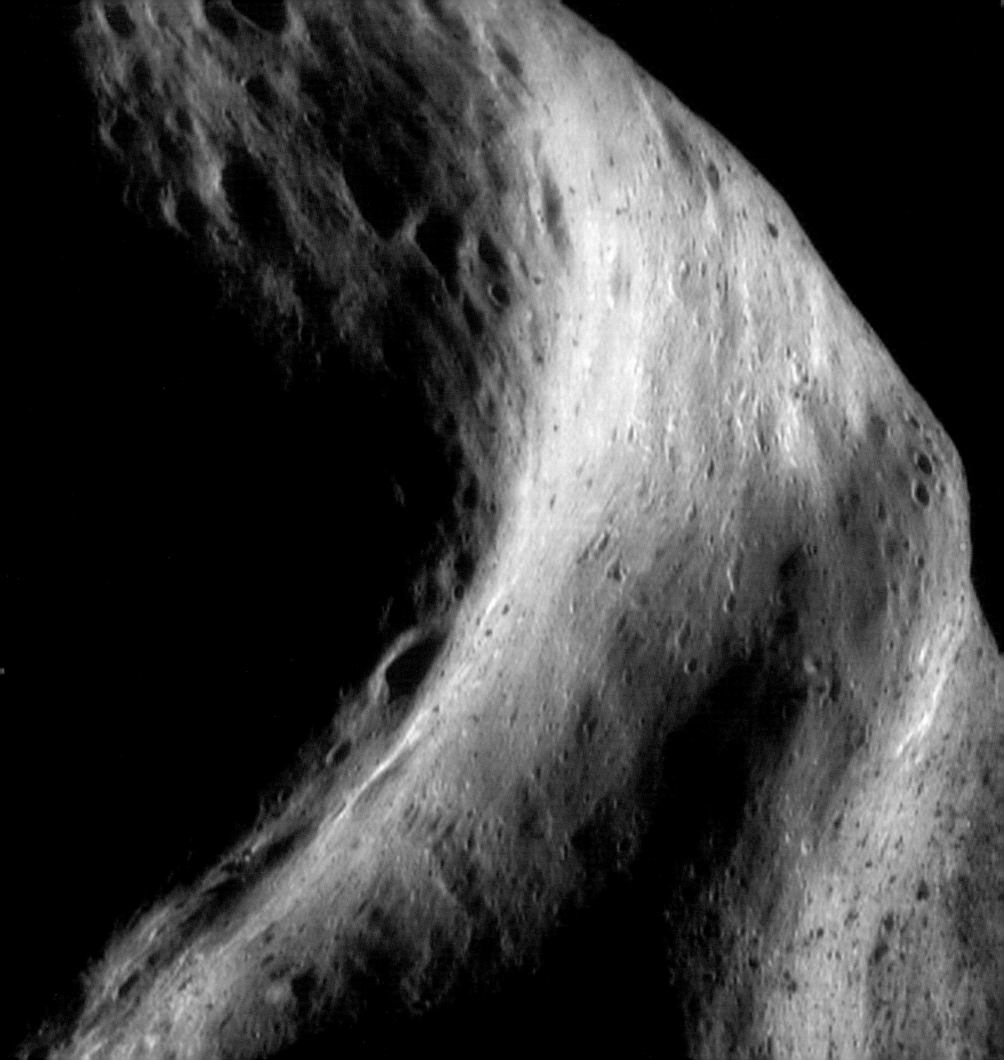

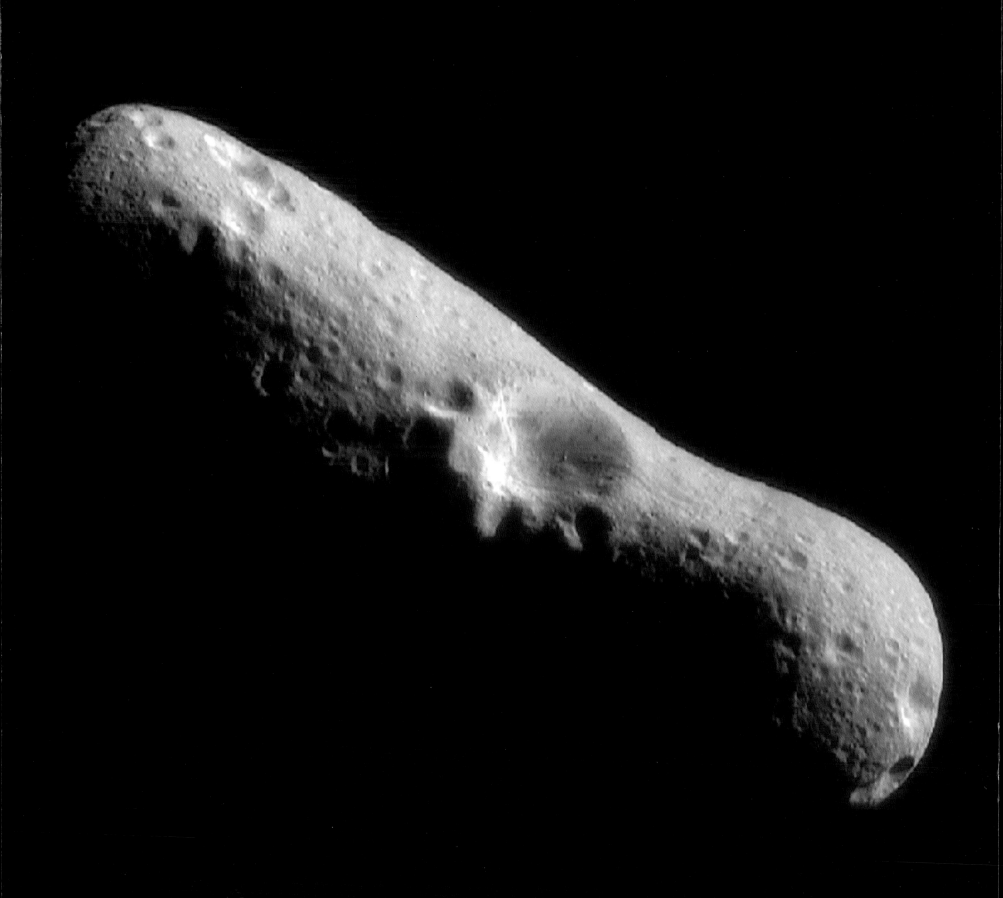

Nothing beats the universe. Nothing is older, larger or more mysterious than the immeasurable cosmos we inhabit and few things have more visual impact than space. The stunning world of stark planetary surfaces, sparkling stars, ghostly nebulae and remote galaxies graces the pages of many books and magazines, and adorns teenage bedroom walls as well as museum exhibits. Images of the universe fire the imagination, and invite us to contemplate our own place in the scheme of things.

Astronomy is often described as the oldest science. Our prehistoric ancestors charted the sky, monitored the motions of the heavens and speculated about the true nature of the sun, the moon, the planets and the stars. In fact, 21st-century astronomers are still doing the same, albeit more precisely and with much more background knowledge, gained over the course of many centuries. And, for the most part, modern astronomy is still very much a visual science, based largely on simply looking.

It is worthwhile noting that, until some four centuries ago, astronomy was exclusively a naked-eye science. Perhaps not everyone possessed the same keen eyesight as the 10th-century Arabic astronomer Abd al-Rahman al-Sufi, who was the first to describe the Andromeda galaxy, or as 16th-century Tycho Brahe, who compiled the most accurate pre-telescopic star catalogue. But, in principle, every human being with two eyes and an inquisitive mind could have made the same discoveries as Aristotle or Copernicus.

The invention of the telescope, around the year 1600, changed all that. Suddenly, a whole new universe was revealed, ready to be observed and explored. Astronomers began to see what had been hidden from view until that moment: mountains on the Moon, dark spots on the Sun, markings on other planets, faint nebulae, glittering star clusters and spiral structure in galaxies. The telescope became the astronomers' artificial eye, providing them with an eagle-eyed view of the distant shores of the cosmos.

But still, for a long time their sight was hampered by the vast distances in the universe and by the blurring effects of the Earth's atmosphere. No matter how large the telescope, there always was a fundamental limit to the smallest detail that you could possibly discern. To overcome this handicap, you had to send your eyes aloft, into Earth orbit or better still, on a voyage of exploration to other worlds. Space flight made this possible, and over the past decades every planet except tiny, distant Pluto has been visited by robotic spacecraft equipped with sensitive cameras, while many dozens of scientific satellites have studied the distant universe from a vantage point high above the troublesome atmosphere of our home planet.

Moreover, astronomers extended their limited vision sideways, by exploring different parts of the electromagnetic spectrum. By studying 'invisible light', like infrared and radio waves, ultraviolet light and X-rays, they discovered a whole new cosmos of violent phenomena and hidden processes. Finally, they were able to hear the whole cosmic symphony instead of just a few notes in a limited frequency range. And nothing could be more rewarding than turning these 'out-of-sight' measurements into images for other people to enjoy. After all, beauty is in the eye of the beholder.

In astronomy, just as in the visual arts, there is good deal of beauty to behold – portraits of intriguing cosmic denizens, serene still lifes, impressive landscapes and an array of abstract compositions. And just as not every painter is in the same league as Van Gogh or Picasso, astronomy has its own grand masters, like NASA's space probe Cassini, the European Very Large Telescope, Mars Global Surveyor and the Hubble Space Telescope. They are among the workhorses of present-day astronomy, and they have provided many of the stunning images on these pages.

Leafing through the book, you might get the impression that the universe is a world of flamboyant colours and surrealistic hues. In many cases, however, considerable artistic freedom is involved, or perhaps more specifically, scientific freedom. The human eye is insensitive to X-rays or radio waves, so these ultra-short and extremely long wavelengths have to be displayed in pseudo-colour. In addition, the science instruments aboard planetary probes and space telescopes are very different from consumer digital cameras: usually, they take black-and-white images at a few very specific wavelengths, which then have to be combined into a 'false colour' composite.

Of course, nobody would want an astronomy photo to show merely what the human eye sees. Suppose we could float in space, just a few light-years away from the Eagle Nebula, one of the most beautiful star-forming regions in our Milky Way galaxy. All we would see with our imperfect eyes would be a small scattering of stars, and maybe an extremely faint glow of nebulosity. Colours would be all but absent, since the human retina is very insensitive to colour at low light levels. It really takes the long exposure, high sensitivity and well-considered wavelength choice of an instrument like the Hubble Space Telescope to reveal the Eagle's splendid beauty.

Likewise, the topography of Mars, the dust-enshrouded core of an active galaxy, or the tiny density clumps in the new-born universe would never be visible without the delicate process of turning hard science data into captivating images – numbers and figures into assemblies of colourised pixels that strongly appeal to our visual senses. And it would be misleading to think that these images are just made for public relations purposes and have little scientific value. In fact, even though astronomers also cherish their incomprehensible graphs, diagrams and spectra, they would never want to live without pictures.

So where does this leave non-astronomers? Actually, in a very comfortable position. We are not obliged to draw scientific meaning from the intricate structures and delicate colours in the amazing imagery that astronomy bestows on us. Of course, it is important to know what you are looking at, so please check the captions at the back of the book. But in principle, we can just sit back, relax and enjoy the trip, opening our eyes, our senses and our minds to the marvels of the universe in which we happen to find ourselves. Who needs hallucinogens or enhanced sensory perception? Nothing beats the universe. Treat yourself.

BILDER AUS DEM UNIVERSUM

Nichts ist faszinierender als das Universum. Es gibt nichts, das älter, größer oder mysteriöser wäre als der grenzenlose Kosmos, in dem wir leben, nichts beeindruckt das Auge nachhaltiger als das All. Die spektakuläre Welt der nackten Planetenoberflächen, funkelnden Sterne, geisterhaften Sternennebel und fernen Galaxien kennen wir aus zahllosen Büchern und Zeitschriften, sie schmückt Wände von Teenagerzimmern ebenso wie Museumsausstellungen. Die Bilder aus dem Universum beflügeln die Fantasie und regen an, über unseren Platz in der Ordnung der Dinge nachzudenken.

Oft nennt man die Astronomie die älteste aller Wissenschaften. Schon zu prähistorischen Zeiten kartierten unsere Vorfahren den Himmel, verfolgten die Bewegungen der Gestirne und fragten sich, wie Sonne, Mond, Planeten und Sterne beschaffen sein mochten. Nicht viel anders, nur genauer und mit sehr viel mehr Hintergrundwissen, das die Menschheit im Lauf der Jahrhunderte erwarb, arbei-ten heutige Astronomen. Im Wesentlichen aber ist die moderne Astronomie immer noch eine visuell geprägte Wissenschaft, die größtenteils auf elementarster Anschauung basiert.

Interessanterweise beruhte bis vor etwa 400 Jahren sogar die gesamte Astronomie auf der Wahrnehmung mit dem bloßen Auge. Wohl mag nicht jeder über so scharfe Augen verfügt haben wie der arabische Astronom Abd ar-Rahman al-Sufi aus dem 10. Jahrhundert, der als Erster die Andromeda-Galaxie beschrieb, oder wie Tycho Brahe, der im 16. Jh. mit damals beispielloser Genauigkeit alle Sterne katalogisierte, bevor noch die ersten Teleskope aufkamen. Doch hätte jeder wissungshungrige Mensch mit zwei Augen und ein bisschen Neugier dieselben Entdeckungen machen können wie Aristoteles oder Kopernikus.

All das änderte sich mit der Erfindung des Teleskops etwa um 1600. Auf einen Schlag öffnete sich ein völlig neues Universum der Beobachtung und Forschung. Nun erlangten die Astronomen einen ersten Einblick in all das, was bisher dem Auge verborgen geblieben war: Mondgebirge, Sonnenflecken, Oberflächenmuster auf Planeten, gerade noch erkennbare Nebel, glitzernde Sternhaufen und spiralförmige Galaxien. Das Teleskop wurde zum „verlängerten Auge" der Astronomen, die nun auch die entlegensten Bereiche des Kosmos messerscharf erkennen konnten.

Doch noch lange sollte der astronomische Blick wegen der riesigen Entfernungen im Universum eingeschränkt und durch die Erdatmosphäre getrübt blei-ben. Wie groß auch immer die Teleskope waren, die atmosphärische Trübung begrenzte die Möglichkeit, Details eindeutig zu erfassen. Hierzu musste sich das „Auge" letztendlich in den Orbit erheben, oder noch besser, sich gleich auf Erkundungsreise in andere Welten begeben. Das wurde erst mit der Raumfahrt möglich – in den vergangenen Jahrzehnten besuchten unbemannte Raumsonden mit hochsensiblen Kameras bis auf den winzigen, fernen Pluto jeden Planeten im Sonnensystem, erkundeten Dutzende Forschungssatelliten die unermesslichen Weiten des Universums von ihrer Warte hoch über der störenden Atmosphäre unseres Planeten.

Auch in anderer Hinsicht erweiterten die Astronomen ihren Blick: Sie untersuchten verschiedene Bereiche des elektromagnetischen Spektrums und erhielten mithilfe „unsichtbaren Lichts" wie infraroter und ultravioletter Strahlung, Röntgenstrahlen und Radiowellen Einblick in ganz neue Welten voller gewaltiger Phänomene und nie zuvor beobachtete Vorgänge. Statt wie bisher nur einzelne Ausschnitte des Kosmos und seiner Phänomene zu betrachten, konnten sie sie plötzlich in beeindruckendem Zusammenklang erleben. Nichts war natürlich schöner, als nun all diese unsichtbaren Messergebnisse in Bilder umzusetzen und so anderen näher zu bringen. Denn erst im Auge des Betrachtenden entfaltet Schönes seinen ganzen Reiz.

Und in der Astronomie gibt es wie in der Kunst viel Schönes zu sehen. Seien es Bilder von faszinierenden Wesen aus dem All, harmonische Stillleben, beeindruckende kos-mische Szenen oder gänzlich abstrakte Kompositionen – wenngleich nicht jedes so wirkt wie ein Van Gogh oder Picasso, so hat doch auch die Astronomie ihre großen Meister. Die produktivsten Raum-Künstler sind die NASA-Raumsonde Cassini, die Marssonde Global Surveyor, das europäische VLT (Very Large Telescope) und das Hubble-Weltraumteleskop, die zu den Lokomotiven der heutigen Astronomie zählen und viele der Bilder geliefert haben, die man auf den folgenden Seiten bestaunen kann.

Beim Durchblättern könnte man den Eindruck gewinnen, das Universum sei eine Welt flammender Farben mit teils surrealistischen Nuancen. Vielfach ist jedoch so einige künstlerische beziehungsweise wissenschaftliche Freiheit mit im Spiel. Das menschliche Auge reagiert nicht auf Röntgenstrahlen oder Radiowellen, sodass diese ultrakurzen und ultralangen Wellen im Bild mit fiktiven Farben wiedergegeben werden müssen. Außerdem funktionieren die wissenschaftlichen Instrumente in Raumsonden und Weltraumteleskopen völlig anders als gängige Fotokameras. Im Allgemeinen nehmen sie mehrere Schwarzweißbilder mit einigen ausgesuchten Spektralbereichen auf, die im Nachhinein zu „konstruierten" Farbbildern zusammengesetzt werden.

Natürlich würde niemand Weltraumfotos sehen wollen, die nur zeigen, was das menschliche Auge erkennt. Würden wir beispielsweise selbst im All schweben, ein paar Lichtjahre von einer der schönsten Sternbildungszonen der Milchstraße, dem Adlernebel, entfernt, dann könnten wir mit der begrenzten Wahrnehmung unserer Augen nur verstreute Sterne und bestenfalls einen schwach schimmernden Nebelschleier erkennen. Farben dagegen würden wir überhaupt nicht sehen, weil die Netzhaut in den unteren Wellenbereichen des Lichts sehr unsensibel für Farben ist. Nur wenn man lange belichtet, mit hoch empfindlichen Geräten wie dem Hubble-Weltraumteleskop arbeitet und die Spektralbereiche sorgfältig auswählt, wird die faszinierende Schönheit des Adlernebels überhaupt erkennbar.

Ähnlich könnten wir auch die Marsoberfläche, den staubverhüllten Kern einer aktiven Galaxie oder die winzigen Verdichtungen in einem neu entstandenen Universum niemals ohne jenen behutsamen Prozess erkennen, in dem harte wissenschaftliche Daten in packende Bilder umgesetzt werden – Zahlen und Graphen, die als Konstellationen aus bunten Pixeln das Auge intensiv ansprechen. Wer jedoch glaubt, all das diene nur den Public Relations und habe keinerlei wissenschaftlichen Wert, liegt falsch. So sehr Astronomen auch für unverständliche Grafiken, Diagramme und Spektren schwärmen, so ungern verzichten sie auf Bilder.

Was heißt das für uns Nichtastronomen? Als Laien haben wir es einfach und müssen den komplexen Strukturen und nuancenreichen Farbtönen der Bilder, die uns die Astronomie beschert, keine wissenschaftlichen Interpretationen abringen. Dennoch ist es immer schön, zu wissen, was man gerade sieht – dazu genügt ein Blick in die Bildlegenden am Ende des Buchs. Vor allem aber kann man sich entspannt zurückgelehnt auf eine Reise ins All begeben und Augen, Sinne und Verstand den Wundern des Universums öffnen, das wir bewohnen. Warum zu Halluzinogenen greifen, wozu die Sinneswahrnehmung erweitern? – Es gibt doch nichts Faszinierendes als das Universum, und nichts ist schöner, als die Bilder auf sich wirken zu lassen!

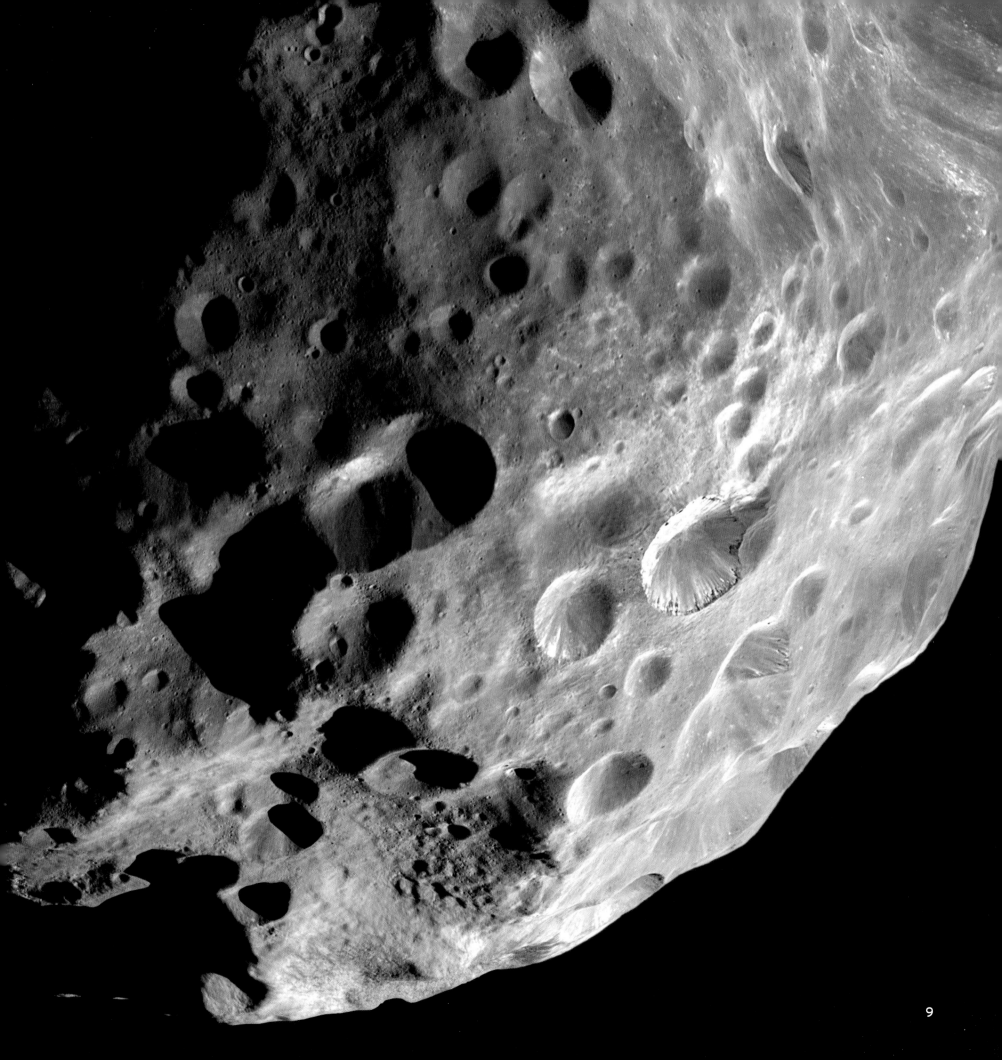

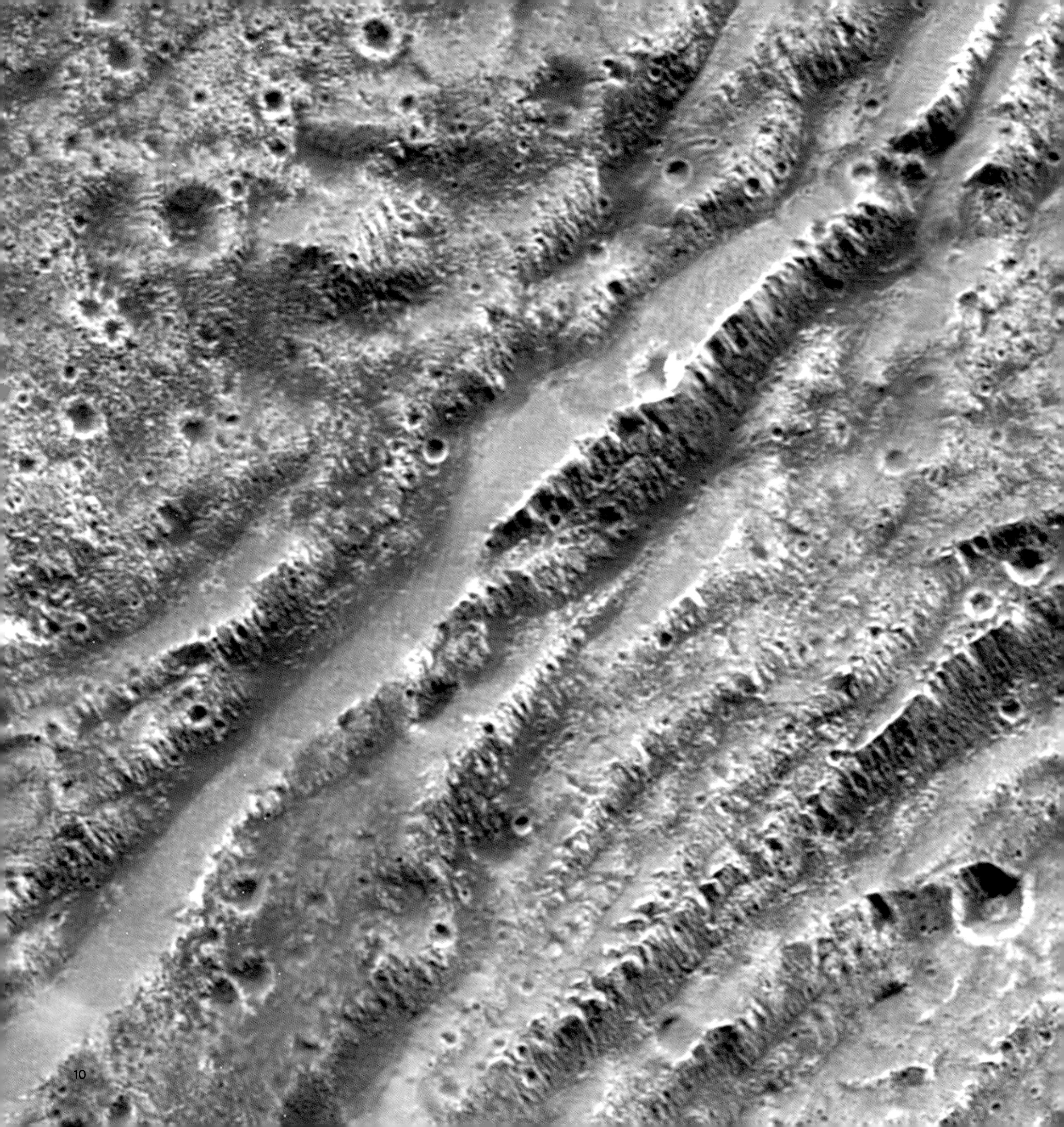

Rien de tel que l'univers. Rien de plus ancien, plus grand ou plus mystérieux que l'incommensurable cosmos dans lequel nous vivons. Et rien, ou presque, ne provoque un plus grand impact visuel que l'espace. Ce monde incroyable, rempli de planètes aux surfaces désolées, d'étoiles scintillantes, de nébuleuses fantomatiques et de galaxies lointaines orne les pages de nos livres et magazines, décore les murs de nos adolescents ou agrémente les expositions de nos musées. Les images de l'univers enflamment l'imagination et invitent à réfléchir à la place de tout un chacun dans le grand ordre des choses.

L'astronomie est souvent décrite comme la plus ancienne des sciences. Nos ancêtres préhistoriques dressèrent des cartes du ciel, étudièrent les mouvements du firmament et s'interrogèrent sur la véritable nature du soleil, de la lune, des planètes et des étoiles. Et les astronomes du 21ème siècle font toujours la même chose, bien que de manière plus précise et sur la base de connaissances beaucoup plus étendues acquises au cours des siècles. Mais pour l'essentiel, l'astronomie moderne reste une science en grande partie visuelle, basée principalement sur ce qui se voit.

Cependant, il est intéressant de noter que, jusqu'il y a environ quatre siècles, l'astronomie était une science qui se pratiquait exclusivement à l'œil nu. Bien entendu, tous n'avaient pas la vue aussi perçante que l'astronome arabe du 10ème siècle Abd al-Rahman al-Sufi, qui fut le premier à décrire la galaxie d'Andromède, ou encore de Tycho Brahé, qui rédigea au 16ème siècle le catalogue d'étoiles pré-télescopique le plus complet au monde ; mais en principe, tout être humain curieux en possession de ses deux yeux et d'un esprit inquisiteur pouvait faire les mêmes découvertes qu'Aristote ou Copernic.

L'invention du télescope, vers 1600, allait tout changer. Un univers entièrement nouveau qui n'attendait que d'être observé et exploré, se révéla soudain à leurs yeux. Les astronomes commencèrent à voir ce qui leur était resté jusqu'alors dissimulé : des montagnes sur la lune, des taches sombres sur le soleil, plusieurs marques sur d'autres planètes, de pâles nébuleuses, des amas d'étoiles étincelantes et des galaxies aux structures en spirales. Le télescope devint l'œil artificiel des astronomes, les dotant d'une vue sans faille pour observer les étendues lointaines du cosmos.

Mais même ainsi, les astronomes ont été encore longtemps gênés par les longues distances propres à l'univers et les « nuages brumeux » causés par l'atmosphère terrestre. Quelle que soit leur taille, tous les télescopes étaient fondamentalement limités quant à la précision des détails discernables. Pour surmonter ce handicap, il fallait porter les yeux au ciel, dans l'orbite de la Terre, ou, mieux encore, s'embarquer pour un voyage d'exploration vers d'autres mondes. Les vols spatiaux ont rendu de tels voyages possibles et au cours des dernières décennies, toutes les planètes, sauf la minuscule et lointaine Pluton, ont été visitées par des engins spatiaux robotisés et équipés de caméras très sensibles, tandis que des dizaines de satellites scientifiques étudiaient le vaste univers depuis une meilleure position, bien au-delà de notre encombrante atmosphère.

Par ailleurs, les astronomes ont étendu les limites de leur champ de vision latéral en explorant différentes zones du spectre électromagnétique. L'étude des « lumières invisibles », comme les ondes infrarouges et radio ou les rayons ultraviolets et les rayons X, leur a permis de découvrir un univers entièrement nouveau, fait de phénomènes violents et de mécanismes cachés. Ils ont fini par être en mesure d'écouter la symphonie cosmique dans sa totalité, et non plus seulement quelques notes d'une plage de fréquences limitées. Et aucune récompense ne peut surpasser celle de transformer ces mesures jusqu'ici mystérieuses en images accessibles à tous. Après tout, la beauté reste quelque chose de subjectif.

En astronomie, comme en art plastique, la beauté ne manque pas à admirer : portraits d'étranges habitants cosmiques, paisibles natures mortes, paysages impressionnants et compositions abstraites en quantités. Et, de même que tous les artistes ne jouent pas dans la même division que Van Gogh ou Picasso, l'astronomie a elle aussi ses grands maîtres, comme par exemple la sonde spatiale Cassini de la NASA, le Very Large Telescope des Européens, la mission Mars Global Surveyor ou le télescope spatial Hubble – pour ne citer que quelques uns des outils essentiels de l'astronomie d'aujourd'hui, qui ont fourni beaucoup des images étonnantes qui suivent.

En feuilletant le livre, le lecteur aura peut-être l'impression que l'univers est un monde aux couleurs flamboyantes et aux nuances surréalistes. Cependant, dans de nombreux cas, il faut compter avec une certaine liberté de l'artiste, ou, devrions-nous dire, du scientifique. En effet, l'œil humain est insensible aux rayons X et aux ondes radio, de sorte que ces longueurs d'ondes, respectivement ultracourtes et extrêmement longues, doivent être rendues par des couleurs artificielles. De plus, les instruments scientifiques embarqués dans les sondes planétaires et les télescopes spatiaux sont très différents des appareils-photo numériques du commerce : ils réalisent en général des images en noir et blanc sur quelques longueurs d'ondes très spécifiques, qui doivent ensuite être fusionnées dans un composite de « fausses couleurs ».

Bien sûr, nul n'aurait intérêt à ce qu'une photo astronomique montre ce que l'œil humain peut voir. Supposons que nous flottons dans l'espace à quelques années-lumière de la nébuleuse de l'Aigle, l'une des plus belles régions de formation d'étoiles de notre galaxie, la Voie lactée. Tout ce que notre vue imparfaite nous permettrait de voir seraient quelques étoiles dispersées et peut-être la lueur à peine visible d'une nébulosité. Les couleurs en seraient quasiment absentes, car la rétine humaine est très peu sensible aux couleurs possédant des niveaux lumineux réduits. La longue exposition, la haute sensibilité et la longueur d'ondes sélectionnée d'un instrument comme le télescope spatial Hubble sont nécessaires pour révéler l'incroyable beauté de l'Aigle.

De même, la topographie de Mars ou les minuscules densités de l'univers nouveau-né ne pourraient être visibles sans la délicate technique qui transforme des données scientifiques brutes en images captivantes (les nombres en assemblages de pixels colorés qui éblouissent notre sens visuel). Ne pas croire pour autant que ces images sont uniquement réalisées pour le grand public ou qu'elles n'ont aucune valeur scientifique. En réalité, même si les astronomes sont très attachés à leurs incompréhensibles graphiques, diagrammes et spectres, ils n'existeraient pas sans ces images qui, après tout, valent bien des milliers de mots

Qu'en est-il, alors, des non-astronomes ? Ils se trouvent en fait dans une position des plus confortables, n'étant, eux, pas obligés de trouver un sens scientifique aux structures complexes et aux couleurs délicates de l'imagerie fantastique qui est l'apanage de l'astronomie. Cependant, comme il est toujours bon de savoir ce que l'on regarde, vous trouverez les légendes à fin du livre. Mais ceux qui le souhaitent peuvent aussi s'installer confortablement, se détendre et profiter du voyage en ouvrant grand leurs yeux, leurs sens et leur esprit aux merveilles de l'univers qui se trouve être le nôtre. Pas besoin d'hallucinogènes ou de perception sensorielle exceptionnelle. Rien ne vaut l'univers, alors, faites-vous plaisir !

IMMAGINI DELL'UNIVERSO

Non c'è niente come l'universo. Niente di più antico, grande e misterioso dell'incommensurabile cosmo in cui viviamo, e quasi nulla ha più impatto visivo delle immagini dello spazio. Un mondo incredibile di superfici planetarie desolate, stelle brillanti, nebulose spettrali e galassie remote adorna le pagine di libri e riviste e decora tanto le camerette dei più giovani come le pareti dei musei. Le immagini dell'universo infiammano la nostra fantasia e ci invitano a riflettere sul posto che occupiamo nell'ordine delle cose.

L'astronomia viene spesso definita come la più antica delle scienze. I nostri antenati della preistoria tracciavano carte del cielo, studiavano i moti della volta celeste e si interrogavano sulla vera natura del Sole, della Luna, dei pianeti e delle stelle. A dire il vero, gli astronomi del XXI secolo fanno ancora le stesse cose, anche se con maggior precisione e con conoscenze di fondo più avanzate fondate su molti secoli di studi. In linea di massima, l'astronomia moderna è ancora una scienza prettamente visiva, basata in gran parte su ciò che vediamo.

Bisogna considerare che, fino a circa quattro secoli fa, l'astronomia era una scien-za basata sull'osservazione ad occhio nudo. Forse non tutti godevano di una vista acuta come quella dell'astronomo arabo del X secolo Abd al-Rahmaan al-Sufi, il primo a descrivere la galassia di Andromeda, o come Tycho Brahe, che nel XVI secolo elaborò il più accurato catalogo stellare dell'epoca anteriore alla comparsa del telescopio. Ma, in linea di massima, qualsiasi essere umano dotato di curio-sità, due occhi e brama di sapere avrebbe potuto realizzare le stesse scoperte fatte da Aristotele e Copernico.

Verso l'anno 1600, l'invenzione del telescopio cambiò radicalmente il panorama rivelando improvvisamente un nuovo universo tutto da osservare ed esplorare. Gli astronomi iniziarono a scrutare tutto ciò che fino ad allora gli era rimasto nascosto alla vista: le montagne della Luna, le macchie solari, la superficie di altri pianeti, nebulose diffuse, ammassi stellari brillanti e strutture a spirale nelle galassie. Il telescopio divenne l'occhio artificiale degli astronomi, offrendogli una nitida veduta delle profondità del cosmo.

Tuttavia, la vista degli astronomi fu offuscata a lungo dalle enormi distanze con cui si confrontavano e dalle distorsioni dovute all'atmosfera terrestre. Per quanto fosse grande un telescopio, la visione dei dettagli non poteva andare oltre certi limiti. Per superarli, bisognava proiettare il punto d'osservazione più in alto, nell'orbita terrestre, o ancor meglio imbarcarsi in un viaggio per esplorare nuovi mondi. I voli spaziali lo hanno reso possibile e negli ultimi decenni tutti i pianeti, tranne il minuscolo e distante Plutone, sono stati visitati da veicoli spaziali senza equipaggio dotati di raffinati dispositivi fotografici, mentre varie decine di satelliti scientifici hanno studiato le profondità dell'universo da un punto d'osservazione privilegiato, al di là della problematica atmosfera del pianeta in cui viviamo.

Inoltre, gli astronomi hanno ampliato la loro visuale sotto altri aspetti, esplorando diverse parti dello spettro elettromagnetico. Grazie all'analisi della "luce invisibile", come i raggi infrarossi, le onde radio, la luce ultravioletta e i raggi X, hanno scoperto un cosmo tutto nuovo caratterizzato da fenomeni impetuosi e misteriosi processi. Finalmente sono riusciti ad ascoltare l'intera sinfonia cosmica invece delle poche note di una gamma di frequenza limitata e non c'è niente di più appagante che trasformare queste misurazioni, altrimenti misteriose, in immagini di cui tutti possano godere. Dopotutto la bellezza risiede negli occhi di chi la osserva.

In astronomia, così come nel campo delle arti visive, c'è una quantità enorme di fonti che ci permettono di contemplare la loro bellezza: ritratti di intriganti abitatori del cosmo, quiete nature morte, imponenti paesaggi e un gran numero di composizioni astratte. E così come non tutti gli artisti raggiungono lo stesso livello di Van Gogh o di Picasso, an-

che l'astronomia ha i suoi grandi maestri, come la sonda spaziale Cassini della NASA, il VLT, il grande telescopio europeo, il Mars Global Surveyor e il telescopio spaziale Hubble. Questi sono solo alcuni dei pezzi forti dell'astronomia moderna e proprio loro ci hanno fornito molte delle straordinarie immagini contenute in quest'opera.

Sfogliando il libro, ci si potrebbe fare un'idea dell'universo come di un mondo dai colori sgargianti e dalle tinte surrealistiche. Tuttavia, in molti casi subentra una certa libertà artistica o, per meglio dire, una libertà scientifica. L'occhio umano non capta né i raggi X né le onde radio pertanto queste lunghezze d'onda, cortissime e lunghissime rispettivamente, devono essere rappresentate con co-lori fittizi. Inoltre, gli strumenti scientifici a bordo delle sonde interplanetarie e dei telescopi spaziali sono assai diversi dagli apparecchi digitali in commercio: normalmente catturano immagini in bianco e nero di onde con una gamma di frequenza ristretta e ben determinata che poi dovranno essere trasformate applicando dei colori "falsi".

Chiaramente, nessuno desidera che una fotografia astronomica mostri ciò che si vedrebbe a occhio nudo. Immaginiamo di fluttuare nello spazio, a solo pochi anni luce dalla nebulosa Aquila, una delle regioni più belle con stelle in formazione della nostra galassia, la Via Lattea. Con la nostra vista imperfetta vedremmo solo qualche stella dispersa e forse un timido bagliore nebuloso, in un ambiente quasi monocromatico, giacché la nostra retina è poco sensibile ai colori se esposta a una luminosità ridotta. Per scrutare l'affascinante splendore di questa nebulosa è assolutamente necessario uno strumento come il telescopio spaziale Hubble, uno strumento molto sensibile capace di realizzare lunghe esposizioni e di captare lunghezze d'onda ben determinate.

Allo stesso modo, la topografia di Marte, il nucleo nebuloso di una galassia attiva o i minuscoli ammassi densi dell'universo in espansione non sarebbero mai visibili senza il delicato processo di conversione: da oscuri dati scientifici a immagini meravigliose, da numeri e calcoli alle composizioni di pixel colorati che affascinano i nostri sensi. Ma sarebbe sbagliato credere che tali immagini vengano rea-lizzate solo per essere sfruttate nel campo delle relazioni pubbliche o che abbiano uno scarso valore scientifico. In realtà gli astronomi, malgrado l'amore per i loro grafici, diagrammi e spettrogrammi incomprensibili, non potrebbero fare a meno di queste fotografie che, dopotutto, valgono molto più delle parole.

Ma c'è spazio per chi non si occupa di astronomia? In realtà, un intero universo. Non dobbiamo per forza estrapolare il significato scientifico delle intricate strutture e dei colori delicati che l'immaginario astronomico ci conferisce. Naturalmente fa sempre piacere sapere cos'è che stiamo osservando e, a tale proposito, si possono consultare le didascalie in inglese alla fine del libro. Ma, se vogliamo, possiamo semplicemente sederci, rilassarci e goderci il viaggio, aprendo bene gli occhi, i sensi e la mente alle meraviglie dell'universo in cui ci ritroviamo. E senza bisogno di allucinogeni o di sti-molare le nostre capacità sensoriali: niente di me-glio che l'universo. Sentitevi a casa vostra!

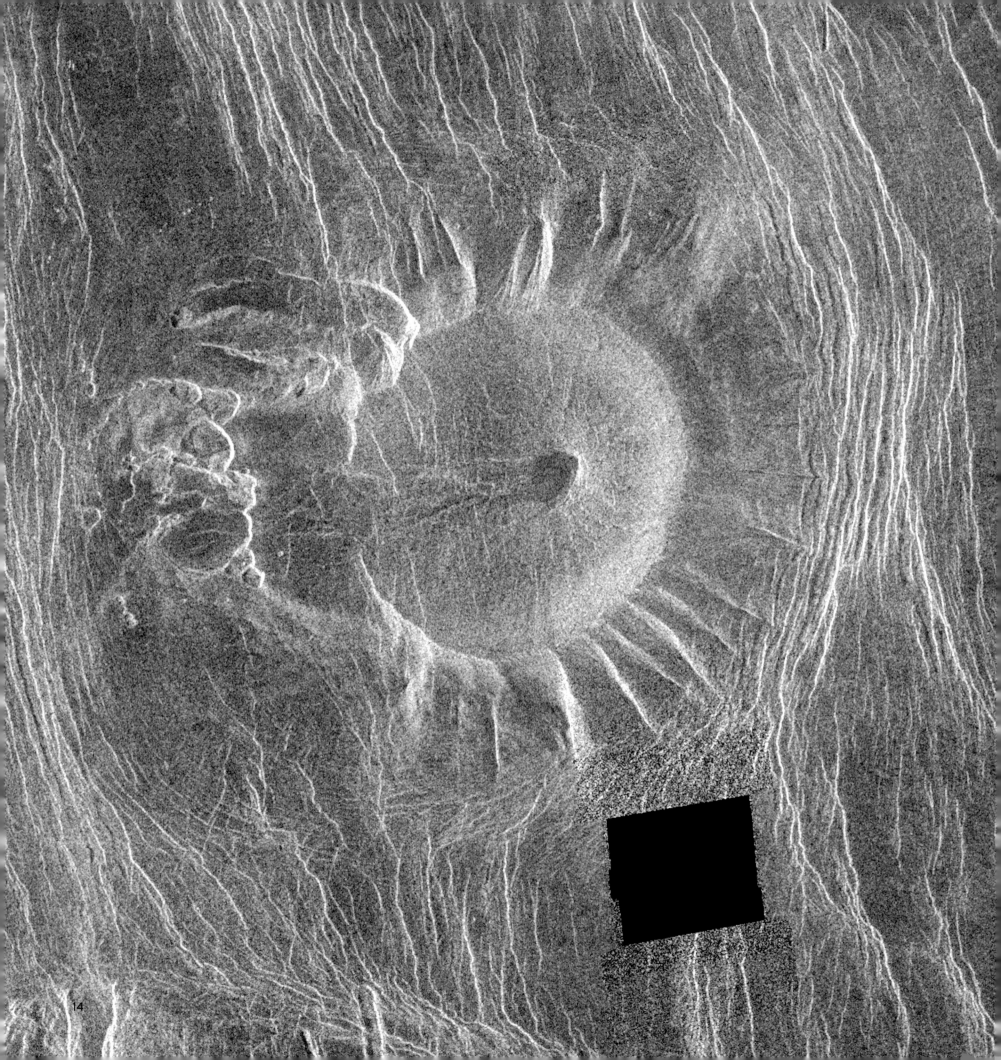

IMÁGENES DEL UNIVERSO

No hay nada como el universo. Nada es más antiguo, más inabarcable o más misterioso que el inconmensurable cosmos en el que habitamos. Y en pocas ocasiones podemos observar imágenes tan impactantes como las del espacio. Un mundo increíble de superficies planetarias inhóspitas, de estrellas brillantes, de nebulosas fantasmagóricas y de galaxias remotas decora las páginas de libros y revistas, y adorna tanto las habitaciones de los adolescentes como las paredes de los museos. Las imágenes del universo disparan la imaginación y nos invitan a reflexionar sobre el lugar que ocupamos en la creación.

La astronomía suele describirse como la ciencia más antigua. Nuestros ancestros prehistóricos trazaron el mapa del cielo, estudiaron sus movimientos y especularon sobre la verdadera naturaleza del Sol, de la Luna, de los planetas y de las estrellas. De hecho, los astrónomos del siglo XXI siguen haciendo lo mismo, aunque con más precisión y un bagaje de conocimientos mucho mayor, recopilado a lo largo de muchos siglos. Por lo demás, la astronomía moderna sigue siendo en gran parte una ciencia visual, basada fundamentalmente en lo que vemos.

Merece la pena destacar que hasta hace aproximadamente cuatro siglos la astronomía era una ciencia que solo contaba con el ojo como instrumento. Tal vez no todo el mundo tuviera una vista tan aguda como el astrónomo árabe del siglo X Abd al-Rahmaan al-Sufi, el primero en describir la galaxia Andrómeda, o como el astrónomo del siglo XVI Tycho Brahe, que recopiló el catálogo de estrellas pretelescópico de mayor precisión. Pero, en principio, cualquier ser humano curioso dotado de dos ojos y una mente inquisitiva podría haber realizado los mismos descubrimientos que Aristóteles o Copérnico.

La invención del telescopio, que tuvo lugar en torno al año 1600, cambió radicalmente la astronomía. De súbito, todo un nuevo universo se reveló a los hombres, prestándose a ser observado y explorado. Los astrónomos tuvieron la ocasión de observar lo que hasta entonces se había ocultado a su visión: las montañas de la Luna, manchas oscuras en la superficie solar, marcas en otros planetas, nebulosas imprecisas, cúmulos estelares brillantes y estructuras espirales que formaban galaxias. El telescopio se convirtió así en el ojo artificial de los astrónomos, ofreciéndoles una nítida panorámica de los confines del cosmos.

Pese a ello, durante largo tiempo la visión de los astrónomos se vio dificultada por las inmensas distancias del universo y las distorsiones debidas a la atmósfera terrestre. Al margen del tamaño que tuvieran, los telescopios siempre encontraban límites en lo que a la precisión de los detalles se refiere. Para solventar este obstáculo, había que proyectar el punto de observación a un lugar más alto, en la órbita terrestre, o, mejor aún, embarcarse en un viaje de exploración hacia otros mundos. Los vuelos espaciales hicieron que este sueño se tornara realidad y, durante las últimas décadas, todos los planetas, salvo el distante y diminuto Plutón, han sido visitados por naves espaciales robóticas equipadas con cámaras de alta sensibilidad, al tiempo que numerosos satélites científicos han estudiado el universo distante desde un punto de vista privilegiado, situado por encima de la problemática atmósfera de nuestro planeta.

Por otro lado, los astrónomos ampliaron su visión lateral, hasta entonces limitada, explorando las distintas partes del espectro electromagnético. Sus estudios de la «luz invisible», como las ondas radioeléctricas o infrarrojas, de la luz ultravioleta y de los rayos X les permitieron descubrir todo un nuevo cosmos de fenómenos violentos y procesos ocultos. Finalmente, fueron capaces de escuchar la sinfonía cósmica completa, en lugar de limitarse a unas cuantas notas de una gama de frecuencias limitada. Y nada podría ser más reconfortante que convertir en imágenes estas mediciones hasta entonces desconocidas para que otras personas pudieran disfrutar de ellas. Al fin y al cabo, la belleza depende del ojo del que mira.

Y en astronomía, como en las artes visuales, hay belleza por doquier: retratos de intrigantes moradores cósmicos, naturalezas muertas que irradian serenidad, paisajes fabulosos y multitud de composiciones abstractas. Y de la misma manera que no todos los artistas están a la altura de Van Gogh o Picasso, la astronomía también cuenta con sus grandes maestros, como la sonda espacial Cassini de la NASA, el nuevo telescopio europeo VLT (así llamado por sus siglas en inglés: Very Large Telescope), el topógrafo global de Marte Mars Global Surveyor y el telescopio espacial Hubble. Estos son solo algunos de los pilares de la astronomía actual, gracias a los cuales se han obtenido muchas de las sorprendentes imágenes de las páginas que siguen.

Es posible que, mientras hojea el libro, se lleve la impresión de que el universo es un mundo de colores estridentes y tonos surrealistas. Sin embargo, en muchos casos esto se debe a cierta libertad artística o, mejor dicho, a cierta libertad científica. El ojo humano no capta ni los rayos X ni las ondas radioeléctricas, por lo que estas longitudes de onda ultracortas y extremadamente largas tienen que representarse con colores ficticios. A ello hay que añadir que los instrumentos científicos que viajan a bordo de las sondas planetarias y los telescopios espaciales son muy distintos de las cámaras digitales de las tiendas: normalmente capturan imágenes en blanco y negro de unas longitudes de onda concretas, que luego tienen que combinarse en un fotomontaje con colores «falsos».

Evidentemente, a nadie le interesaría que una fotografía astronómica mostrara lo que puede observarse a simple vista. Imagine que pudiéramos flotar en el espacio, a tan solo unos cuantos años luz de la Nebulosa del Águila, una de las constelaciones en formación más maravillosas de nuestra galaxia, la Vía Láctea. Lo único que nuestros ojos imperfectos nos permitirían ver sería unas cuantas estrellas diseminadas y, tal vez, un levísimo destello de nebulosidad. La ausencia de color sería total, ya que la retina humana no capta el color cuando el nivel de luz es bajo. Es necesario contar con la larga exposición, la alta sensibilidad y la especificación de las longitudes de onda de un instrumento como el telescopio espacial Hubble para admirar la Nebulosa del Águila en todo su esplendor.

De la misma manera, la topografía de Marte, o los cúmulos de densidad diminutos del universo recién nacido no podrían detectarse sin el delicado proceso de convertir enrevesados datos científicos en imágenes cautivadoras, o números y cifras en composiciones de píxeles de colores que atraen irremisiblemente a nuestros sentidos visuales. Pero no se lleve a engaño: estas imágenes no solo se elaboran a modo de promoción y es un error pensar que no tienen valor científico. En realidad, pese a la devoción que profesan los astrónomos por sus incomprensibles gráficos, diagramas y espectros, no renunciarían a estas imágenes, que, a fin de cuentas, valen más que mil palabras.

¿Adónde conduce todo esto a los legos en astronomía? En realidad, a una posición muy cómoda. No estamos obligados a extraer un significado científico de las intricadas estructuras y delicados colores de las impresionantes imágenes que la astronomía nos regala. Evidentemente, resulta gratificante saber qué se está mirando, por lo que les recomiendo que se remitan a los pies de foto en inglés incluidos al final del libro. Pero, si lo deseamos, podemos sentarnos, relajarnos y disfrutar del viaje, manteniendo bien abiertos nuestros ojos, sentidos y mentes para observar las maravillas del universo en el que se nos ha dado vivir. ¿Quién necesita alucinógenos? Nada supera al universo. ¡Dese el placer de comprobarlo!

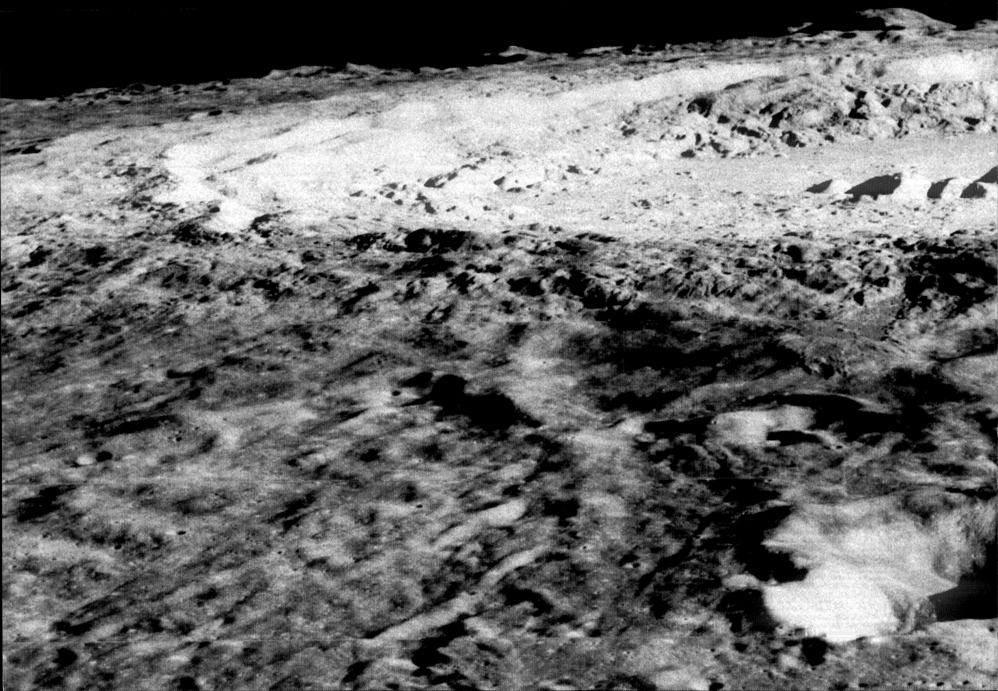

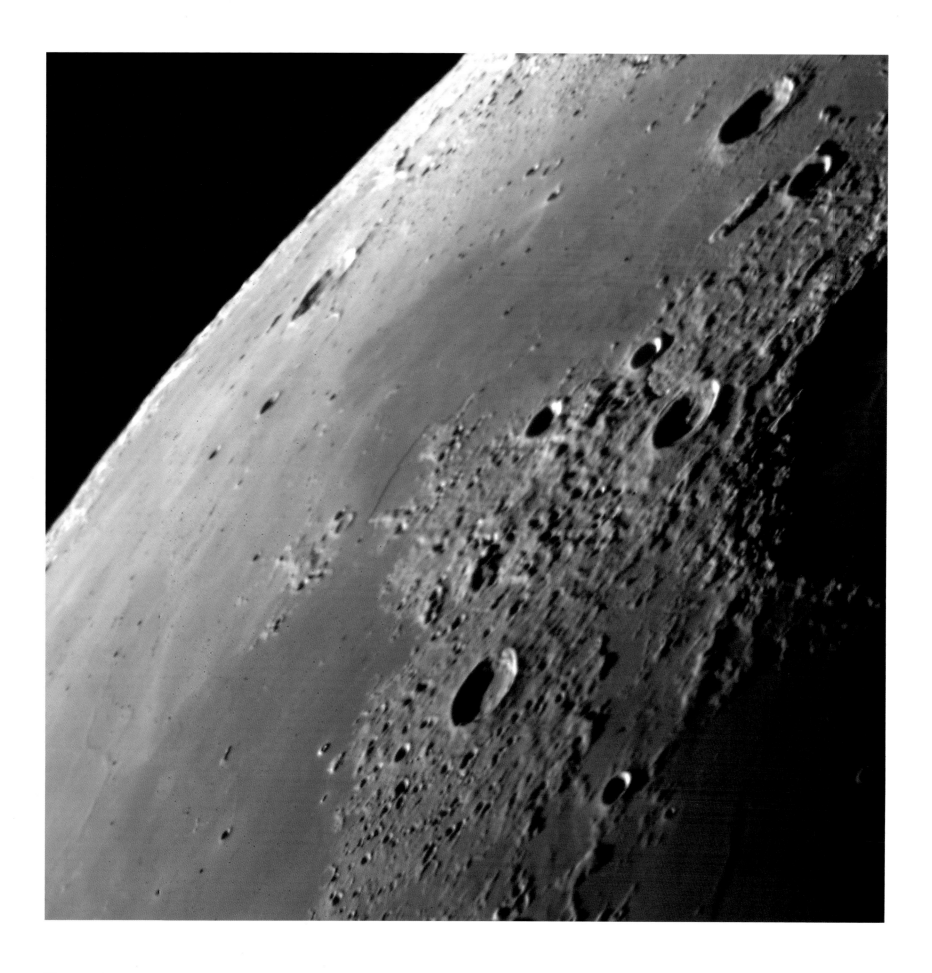

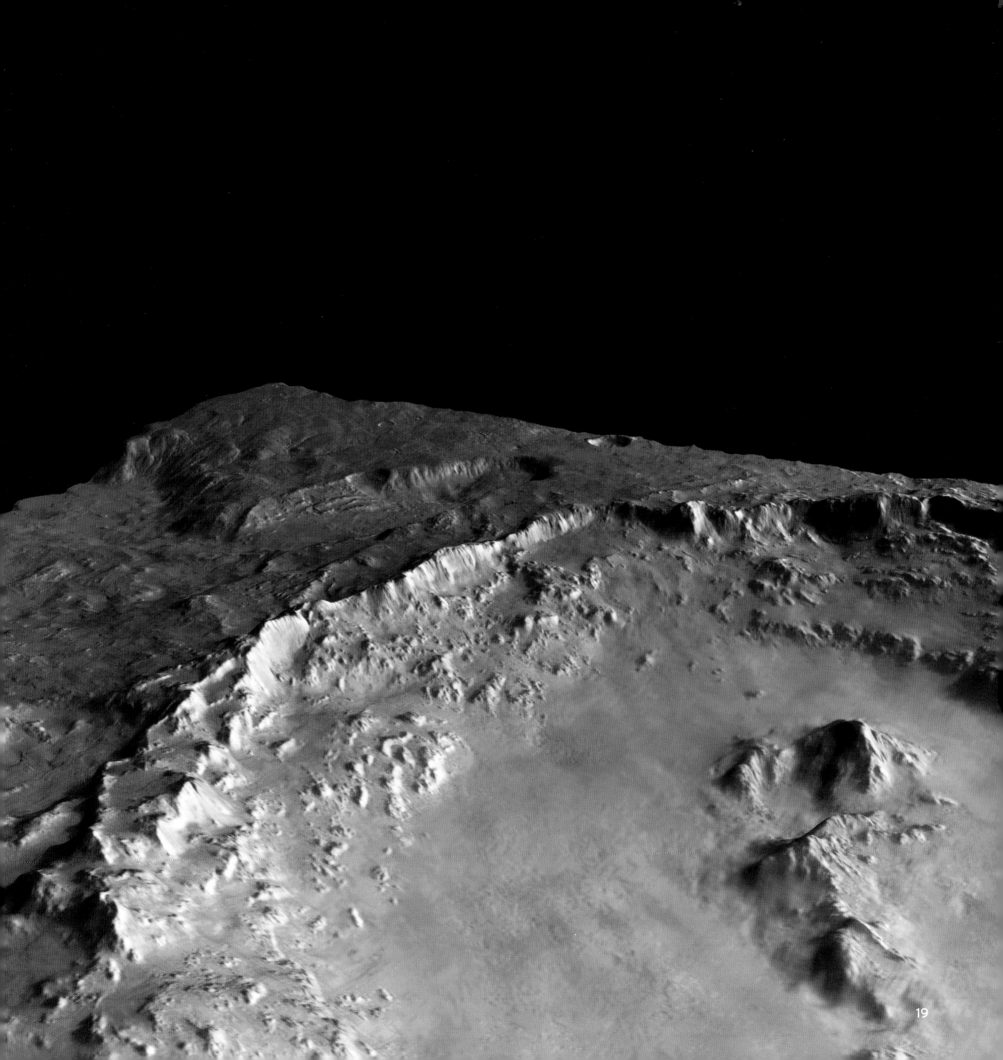

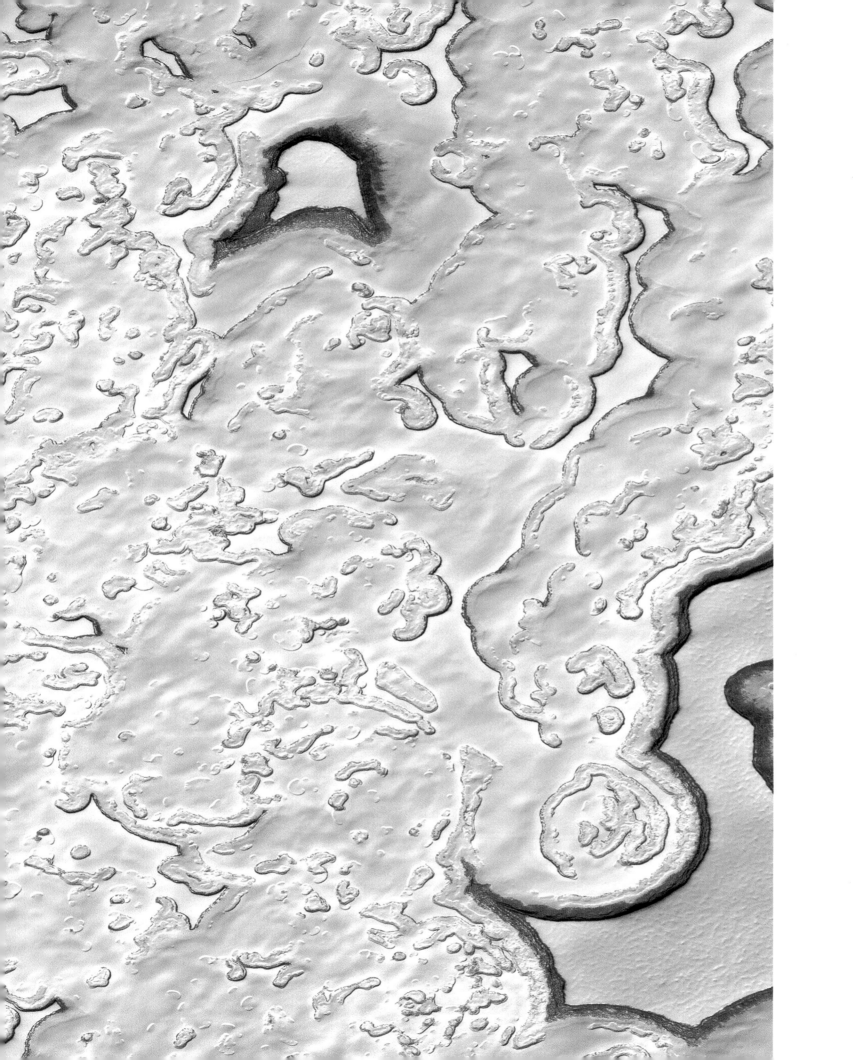

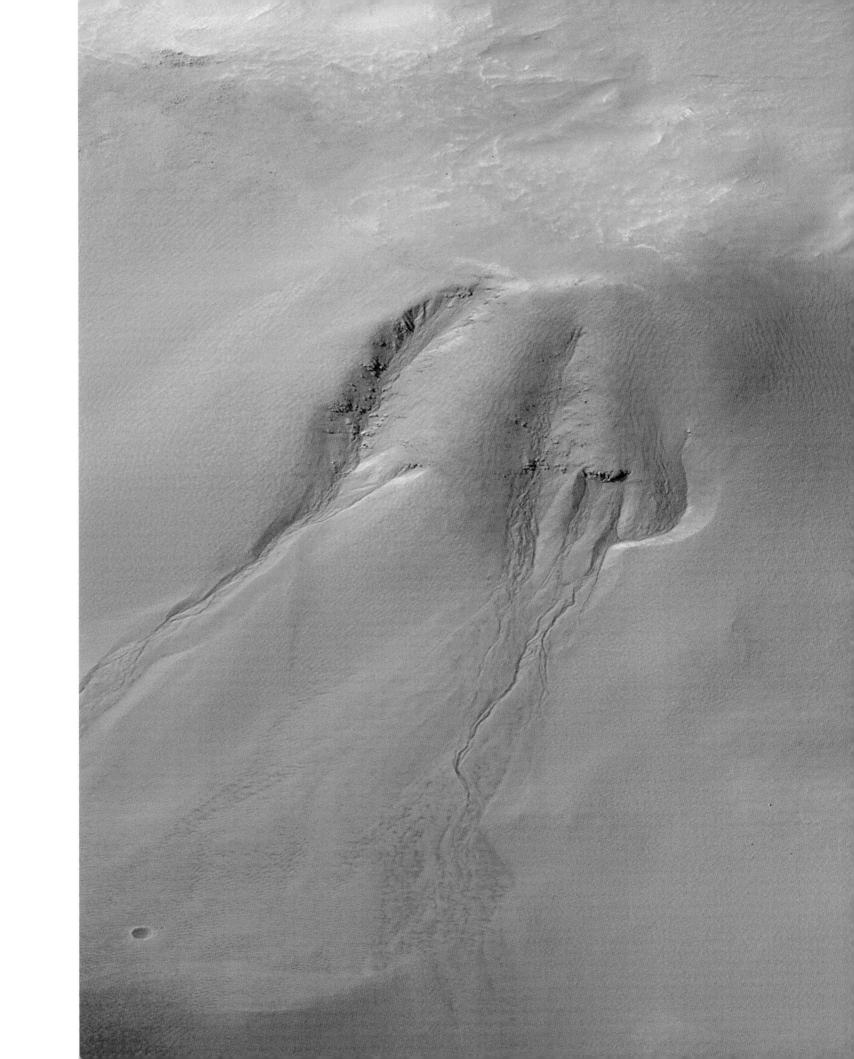

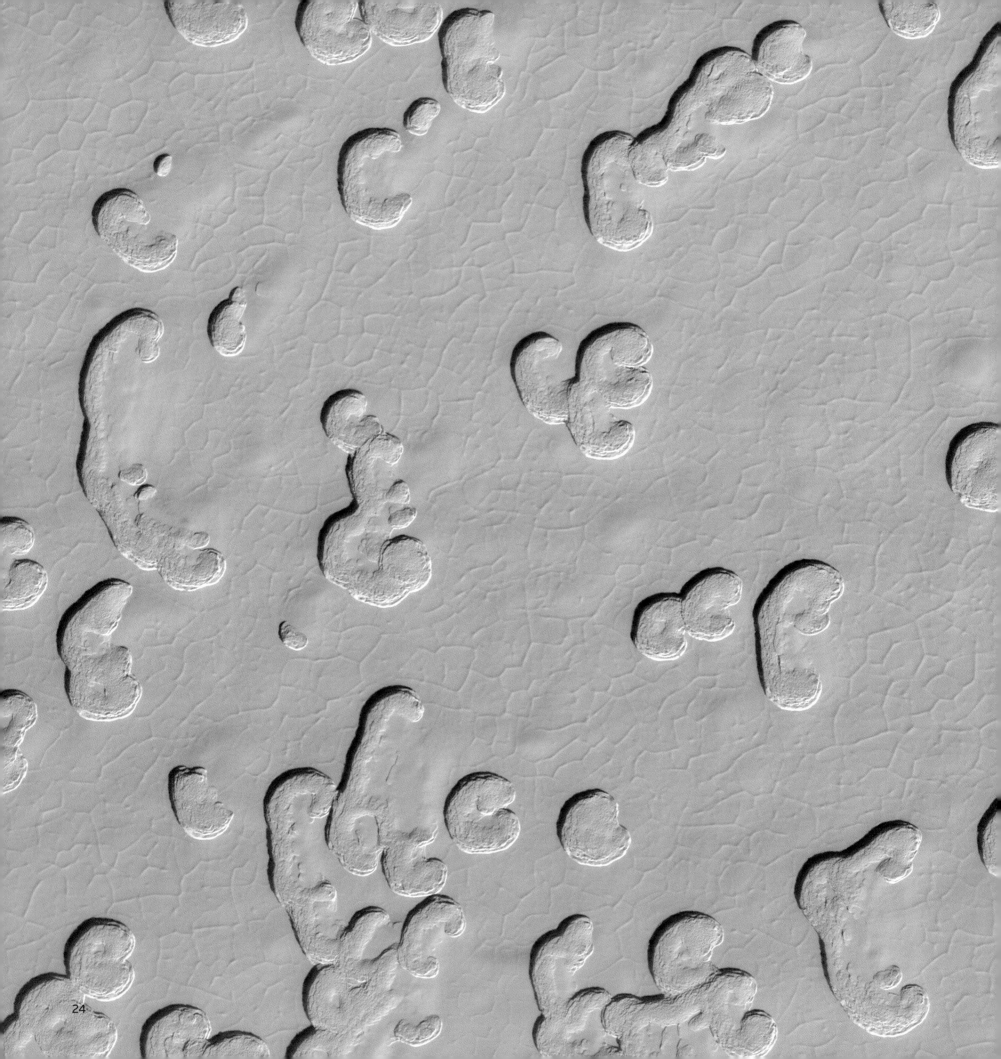

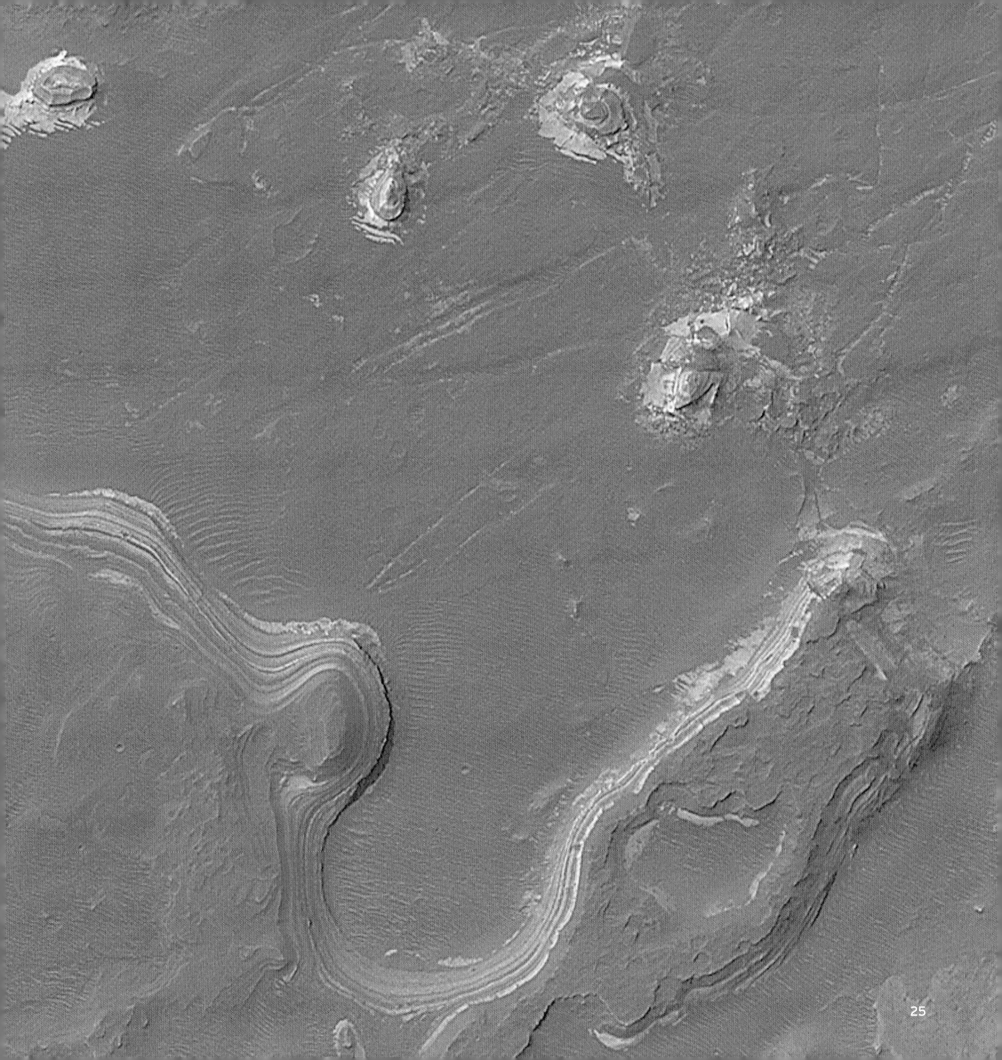

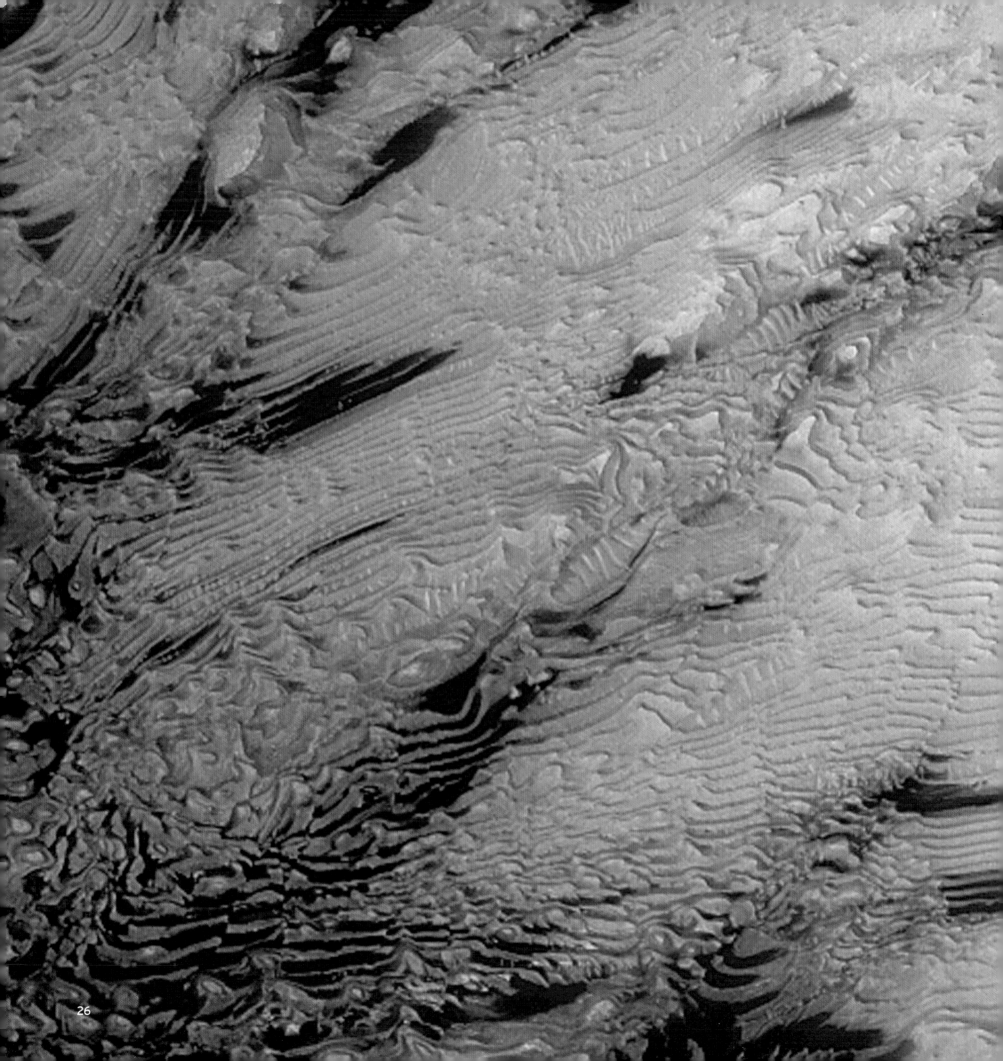

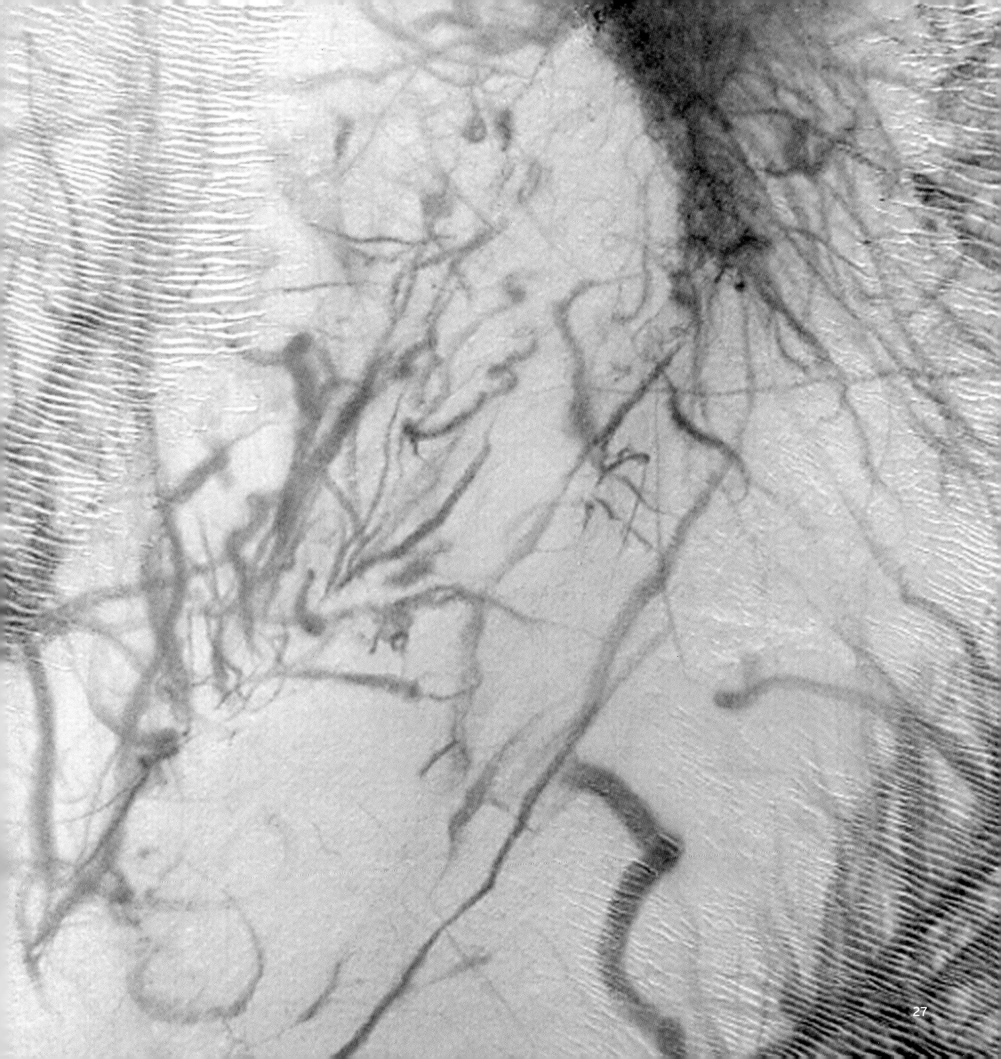

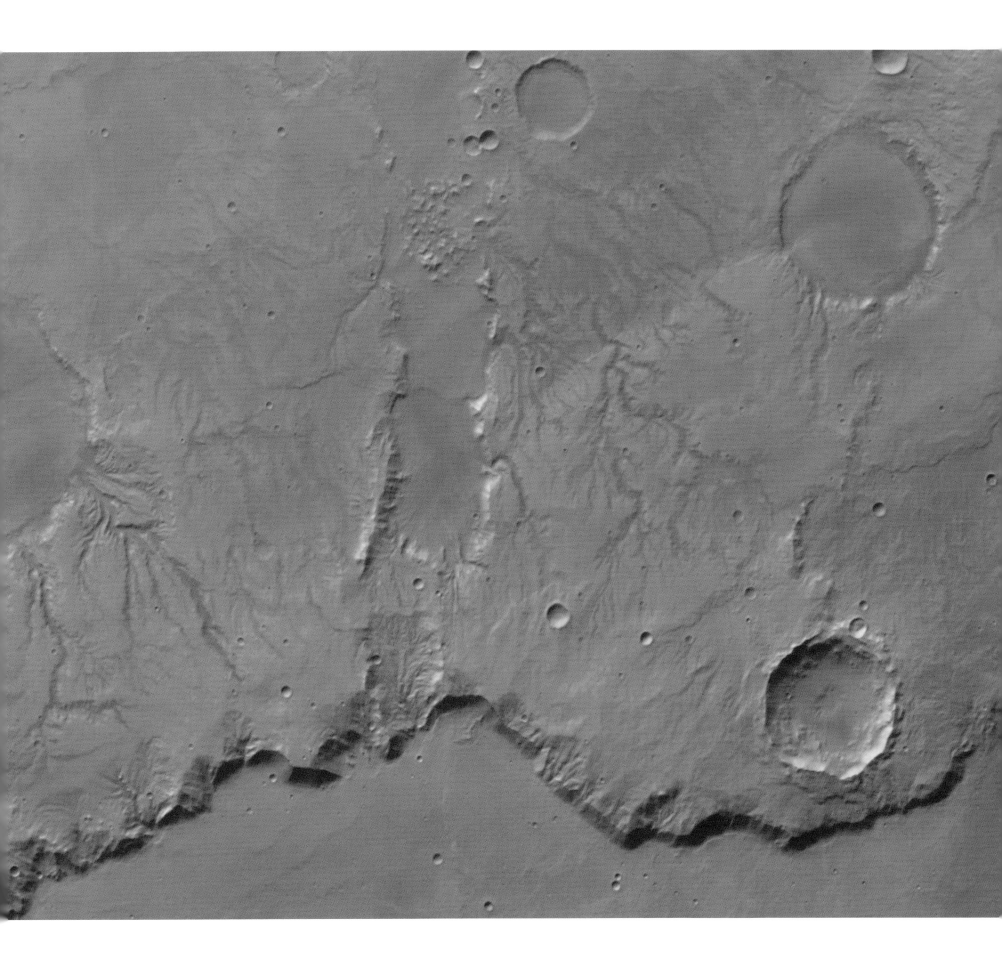

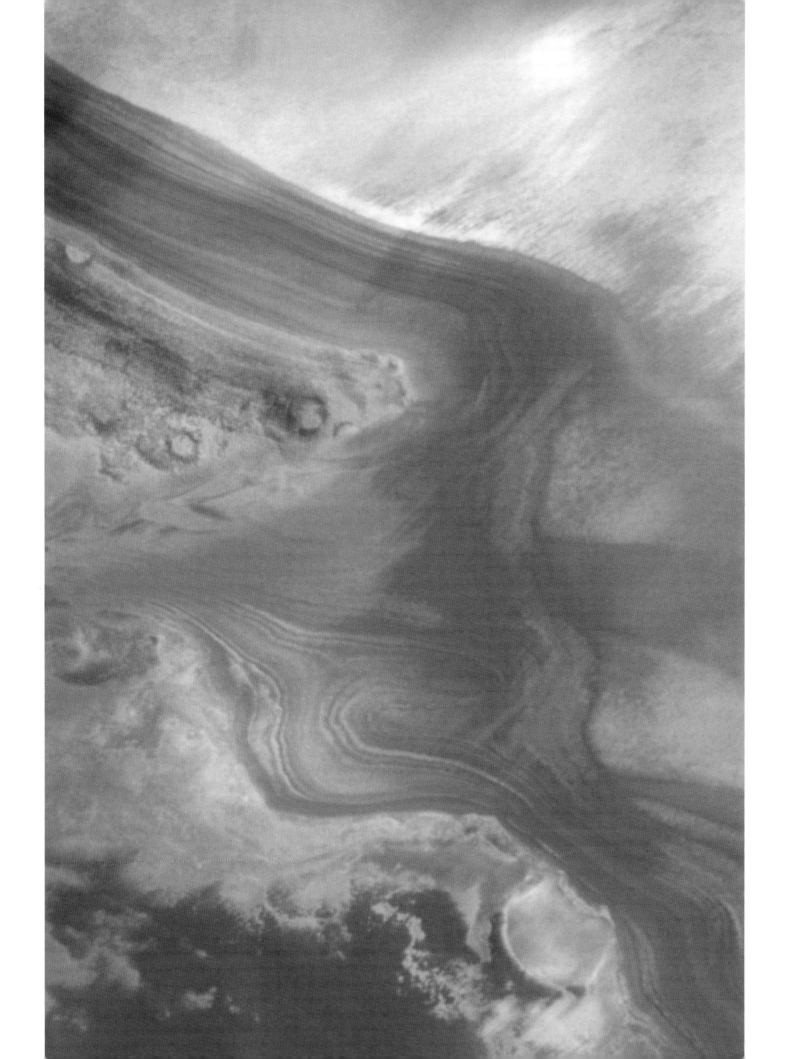

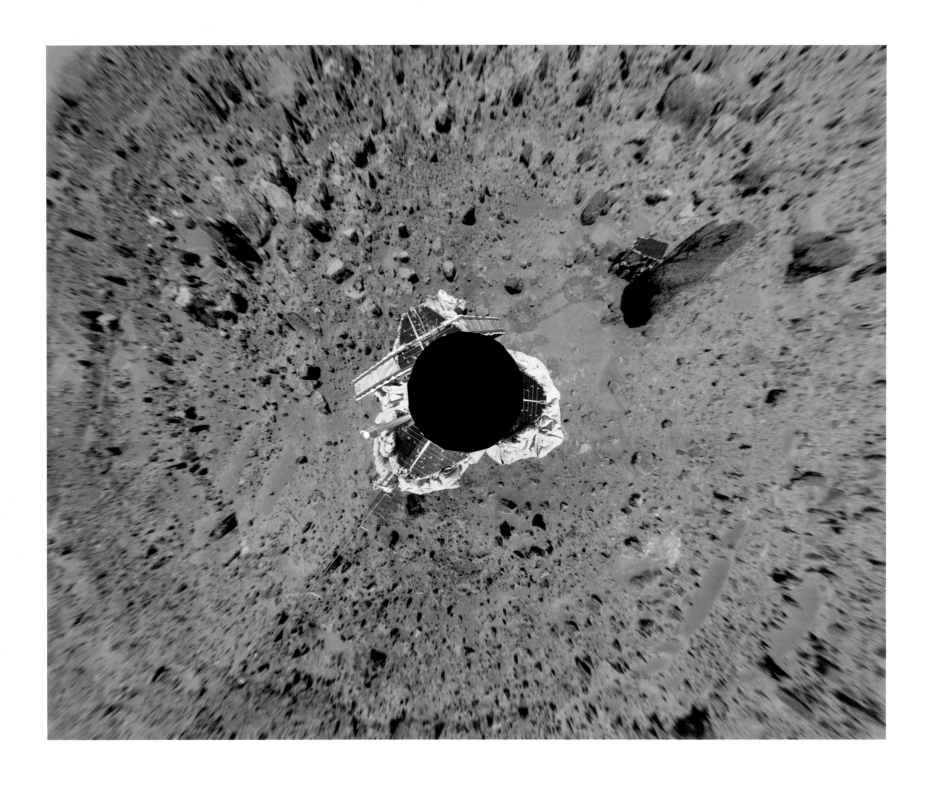

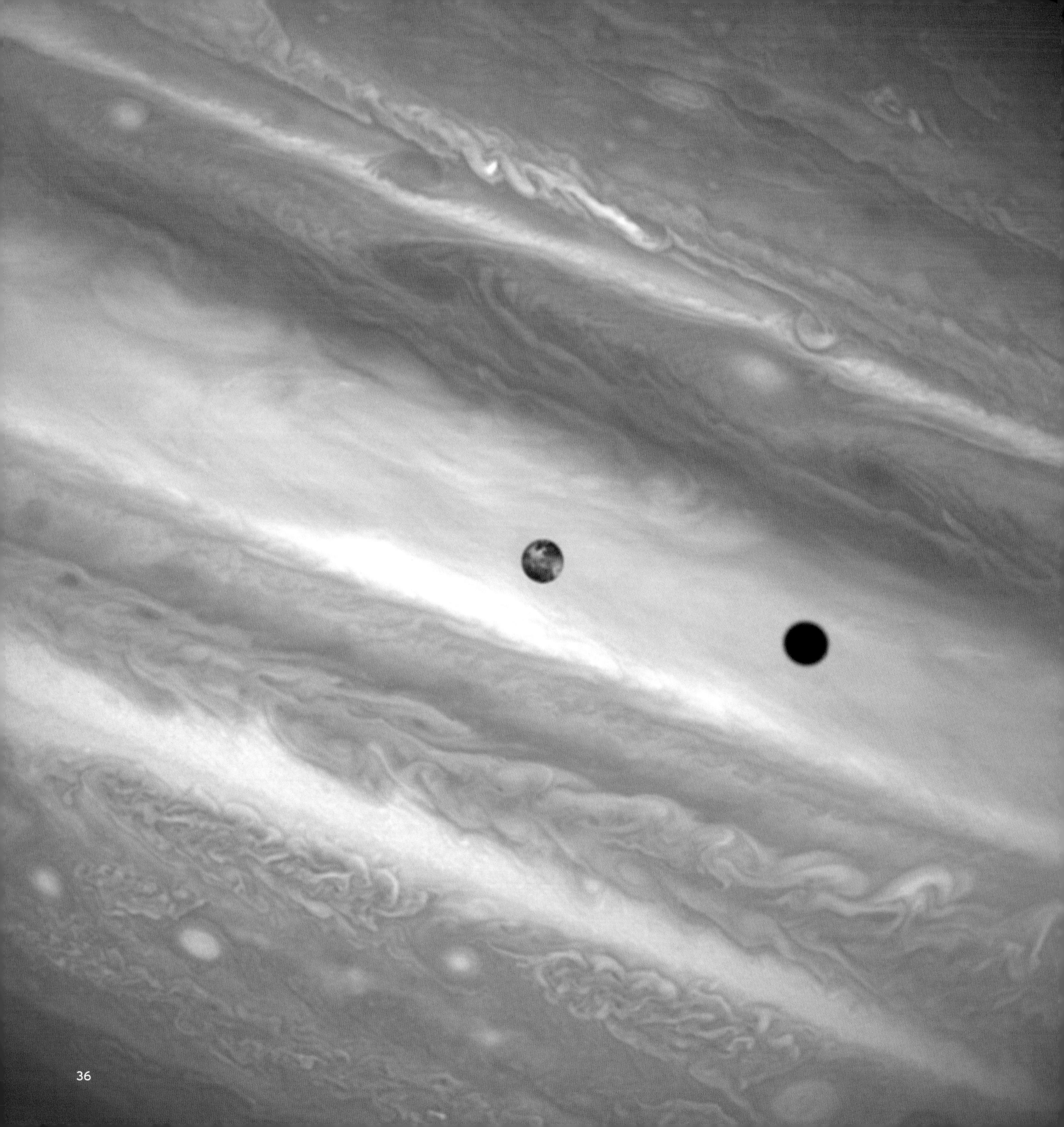

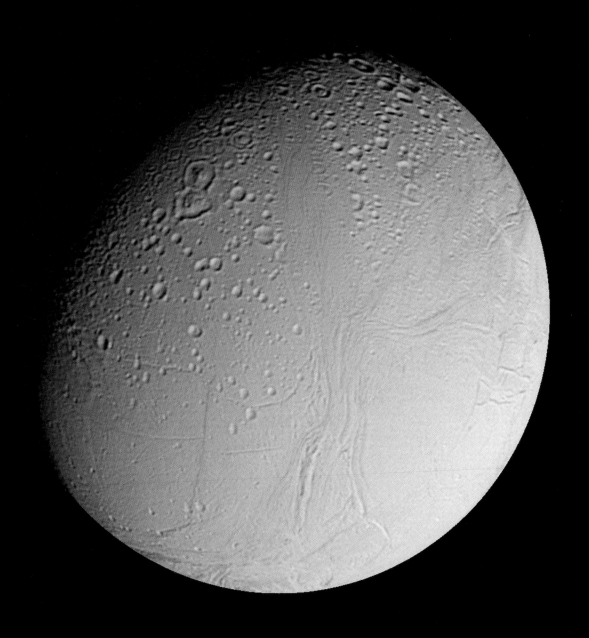

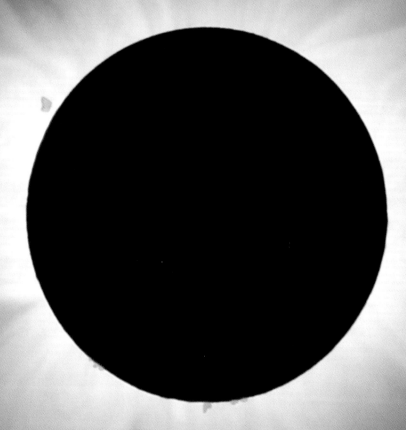

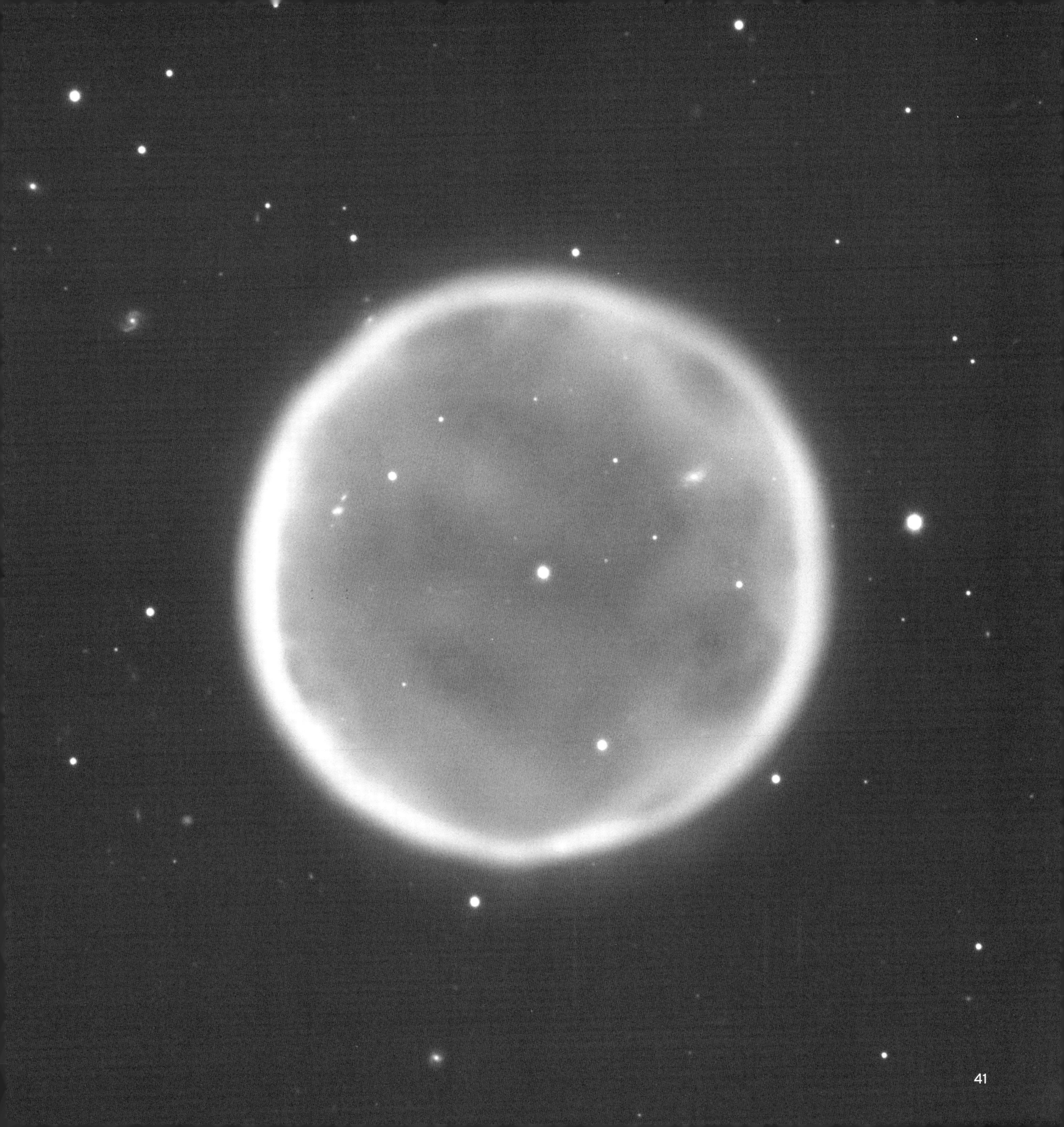

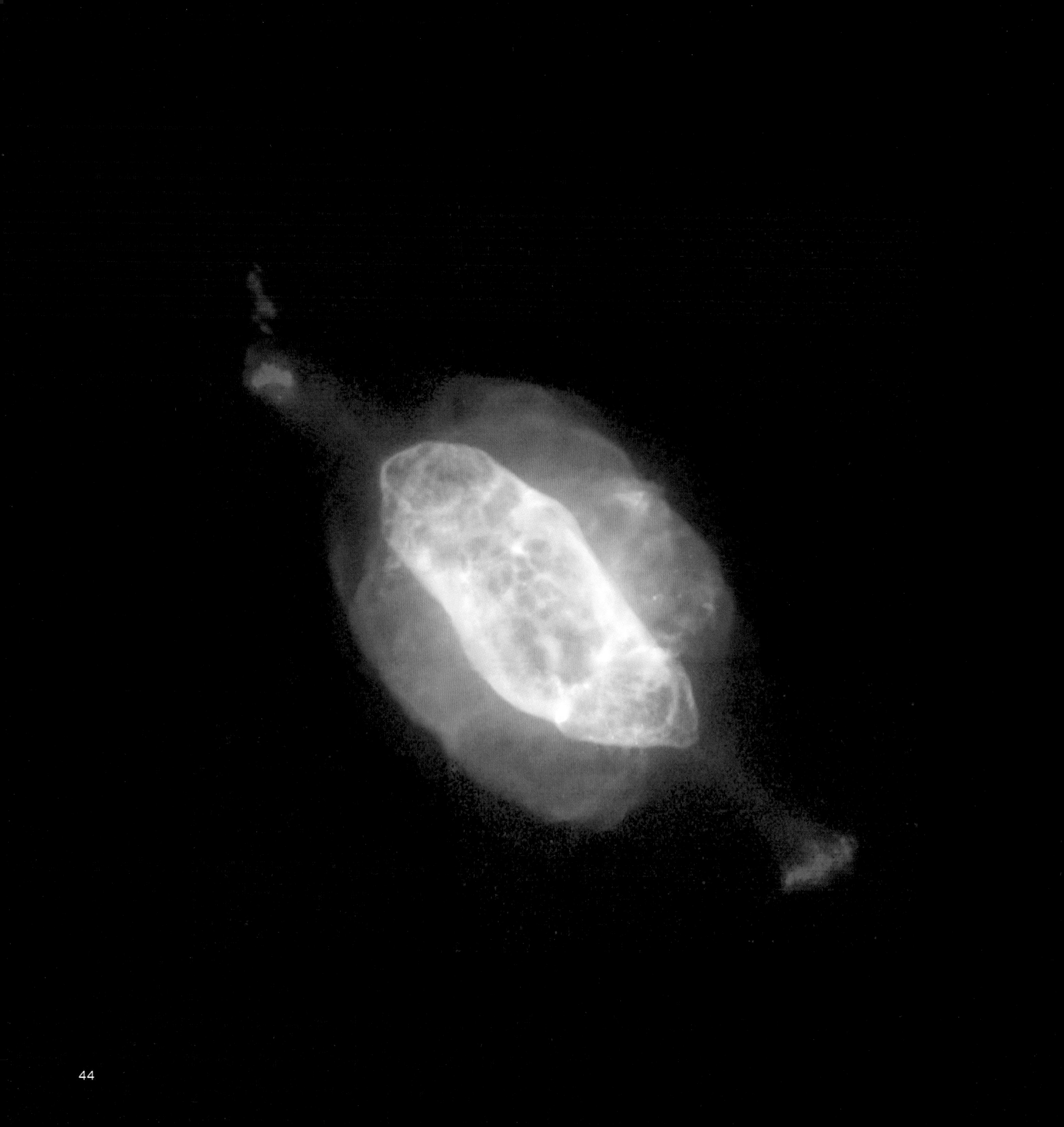

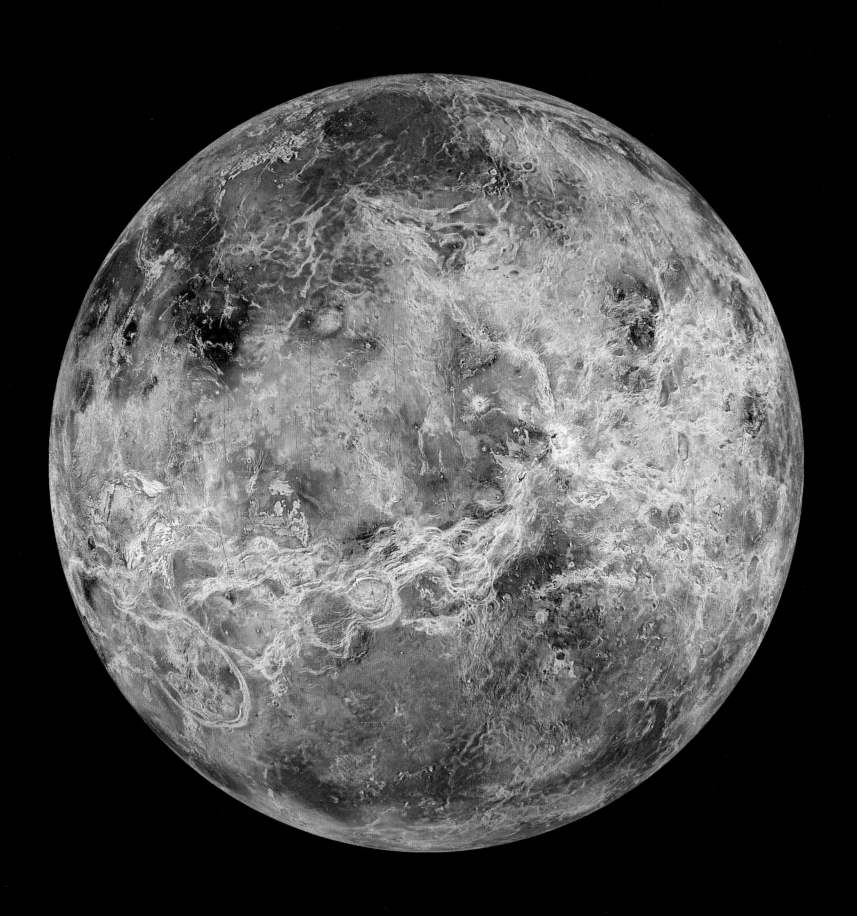

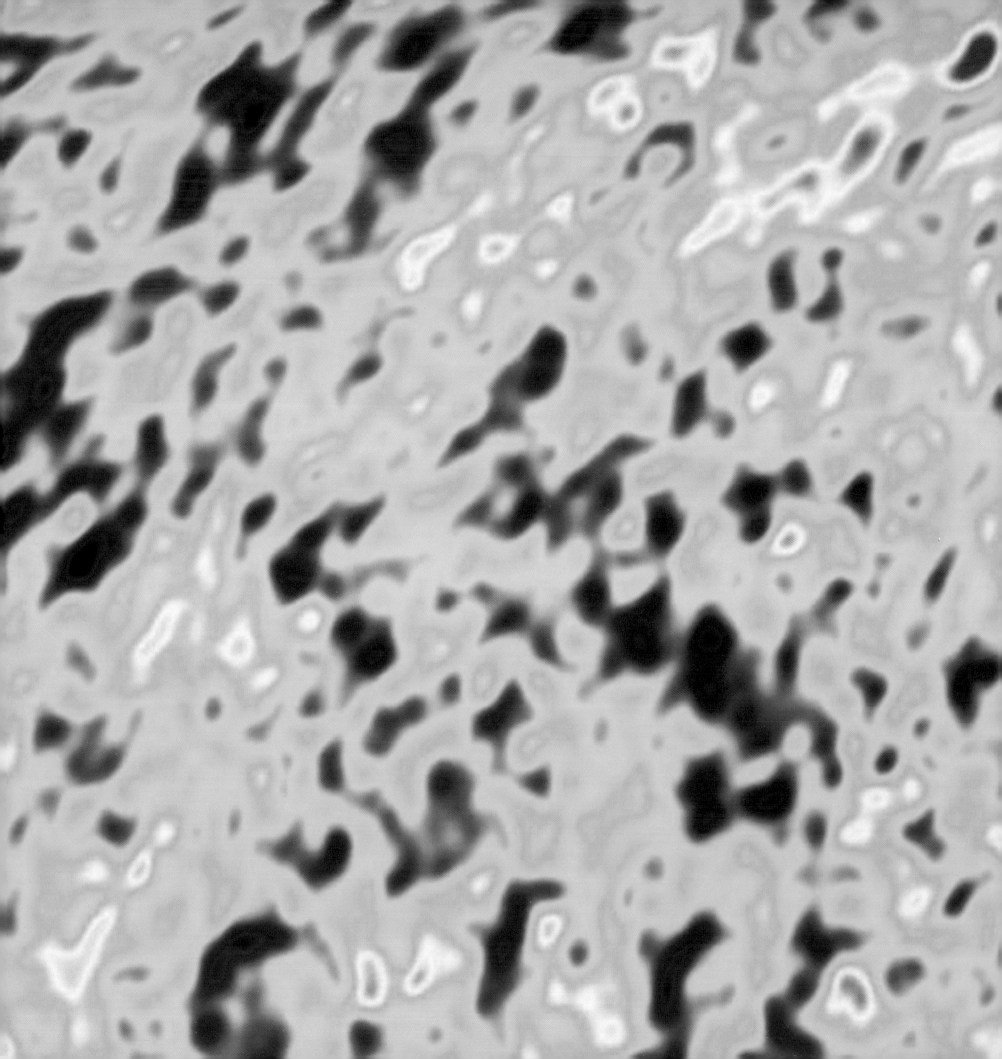

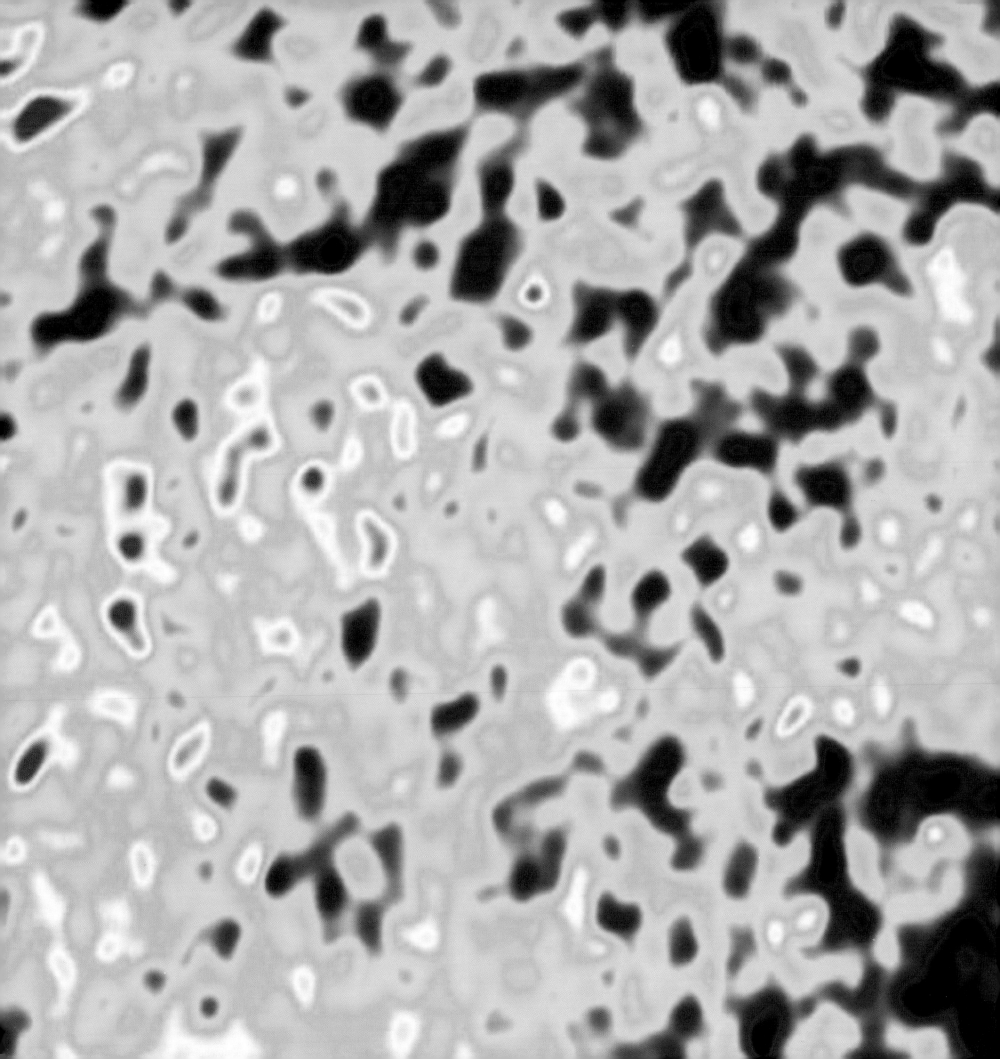

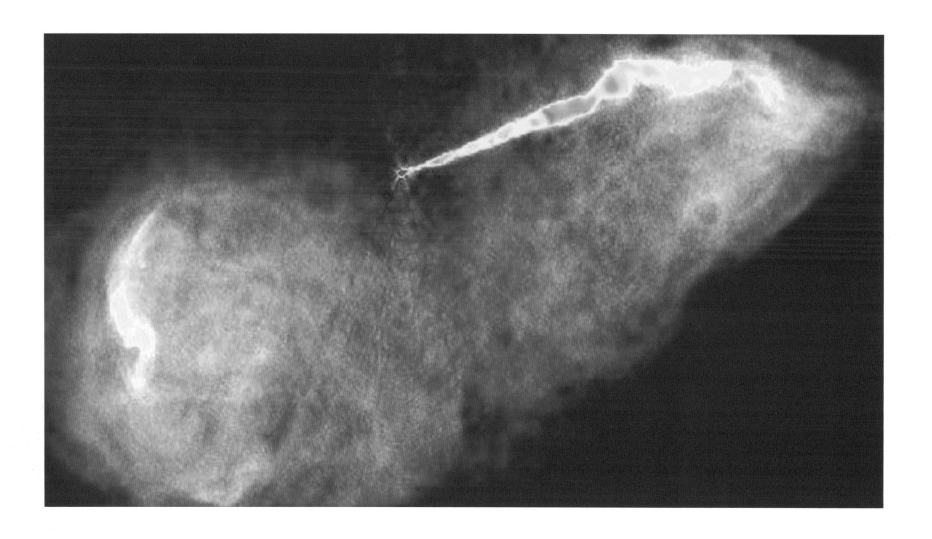

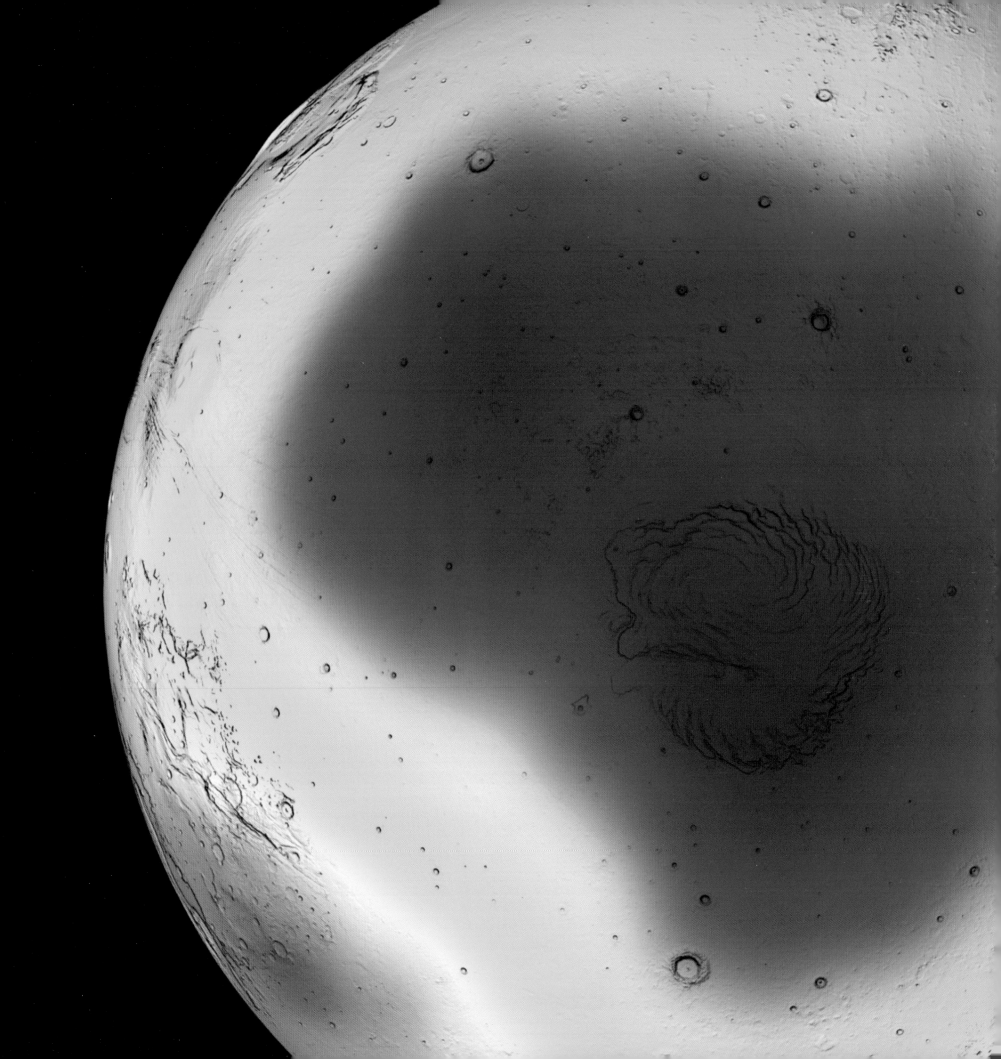

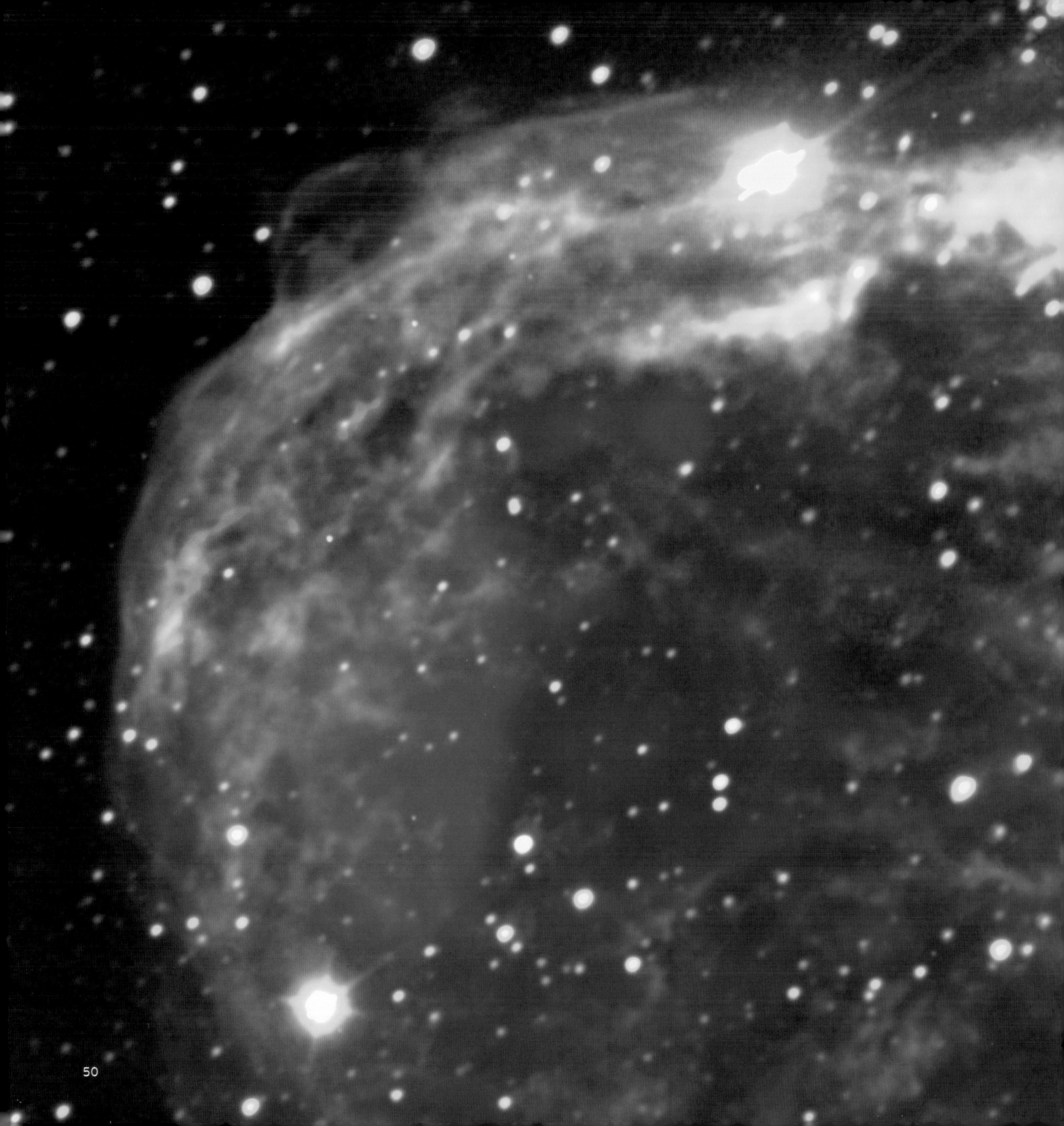

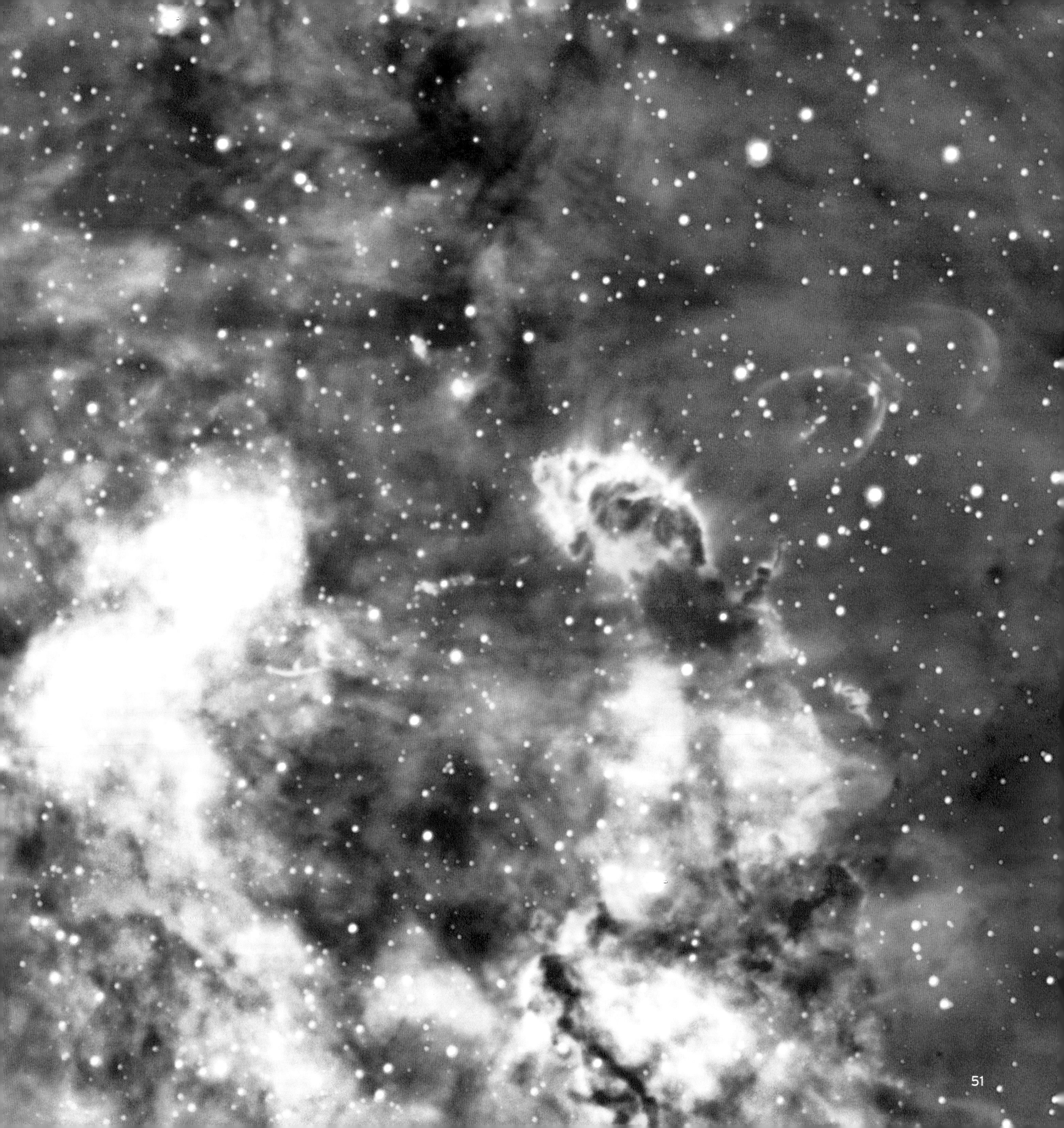

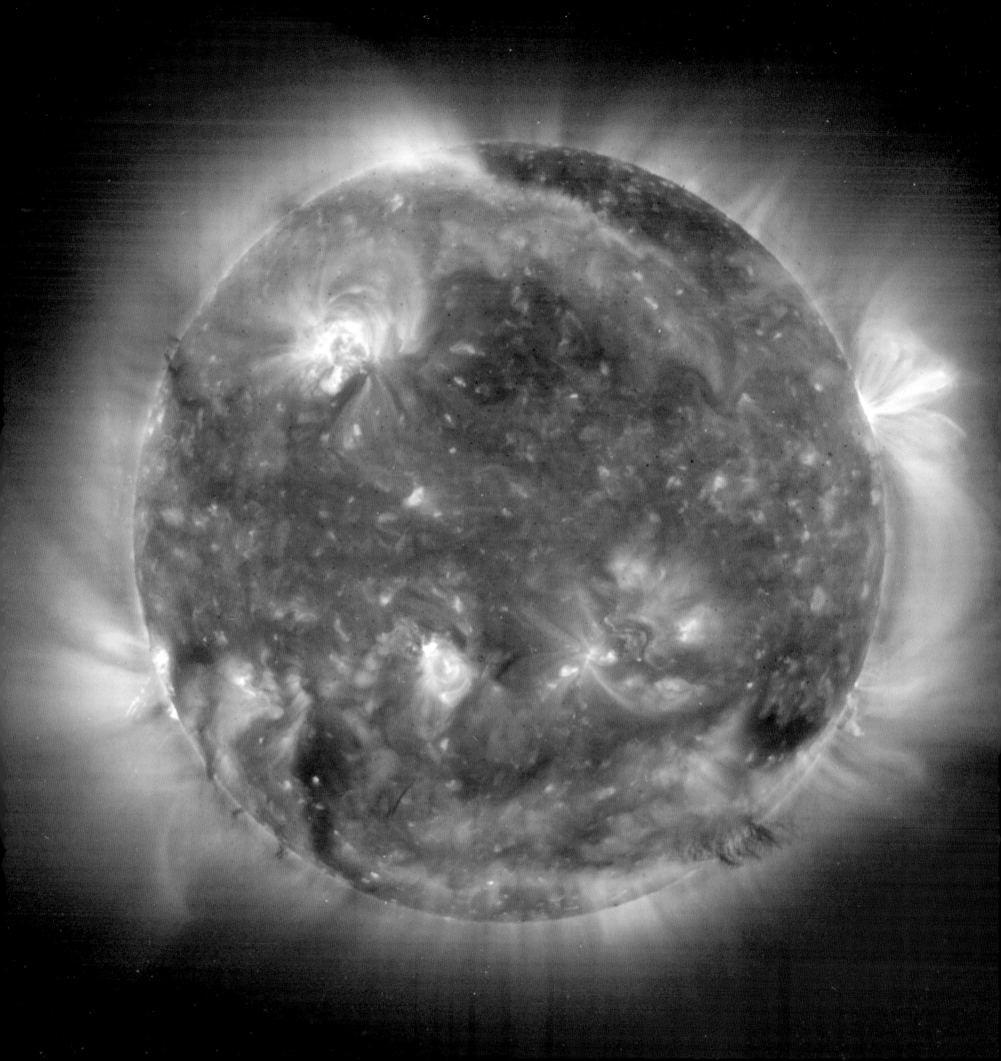

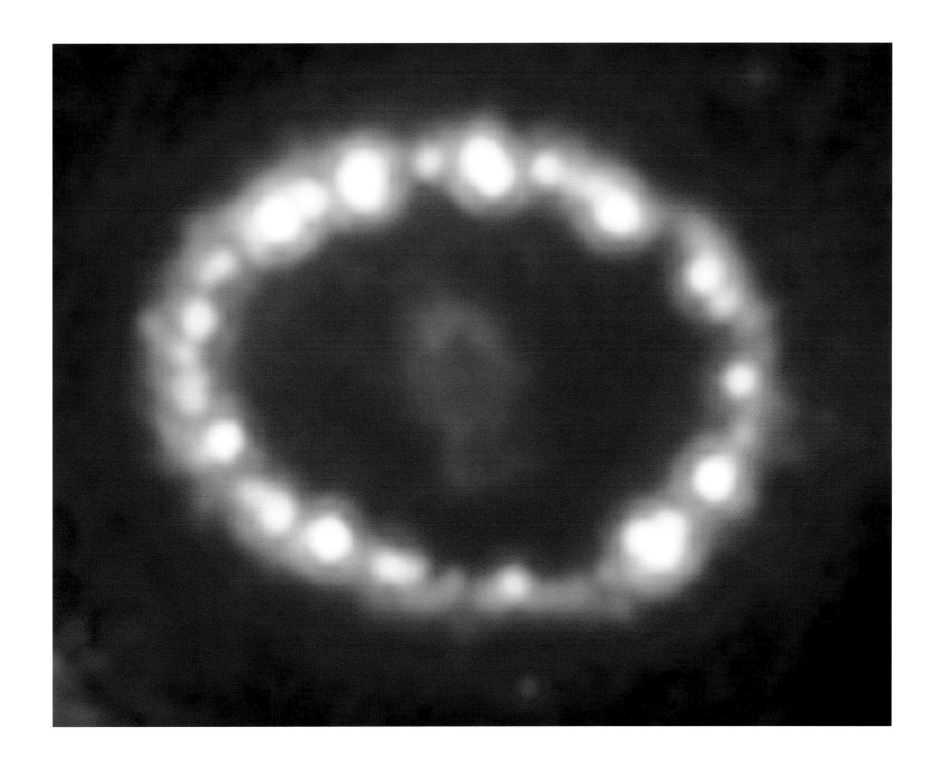

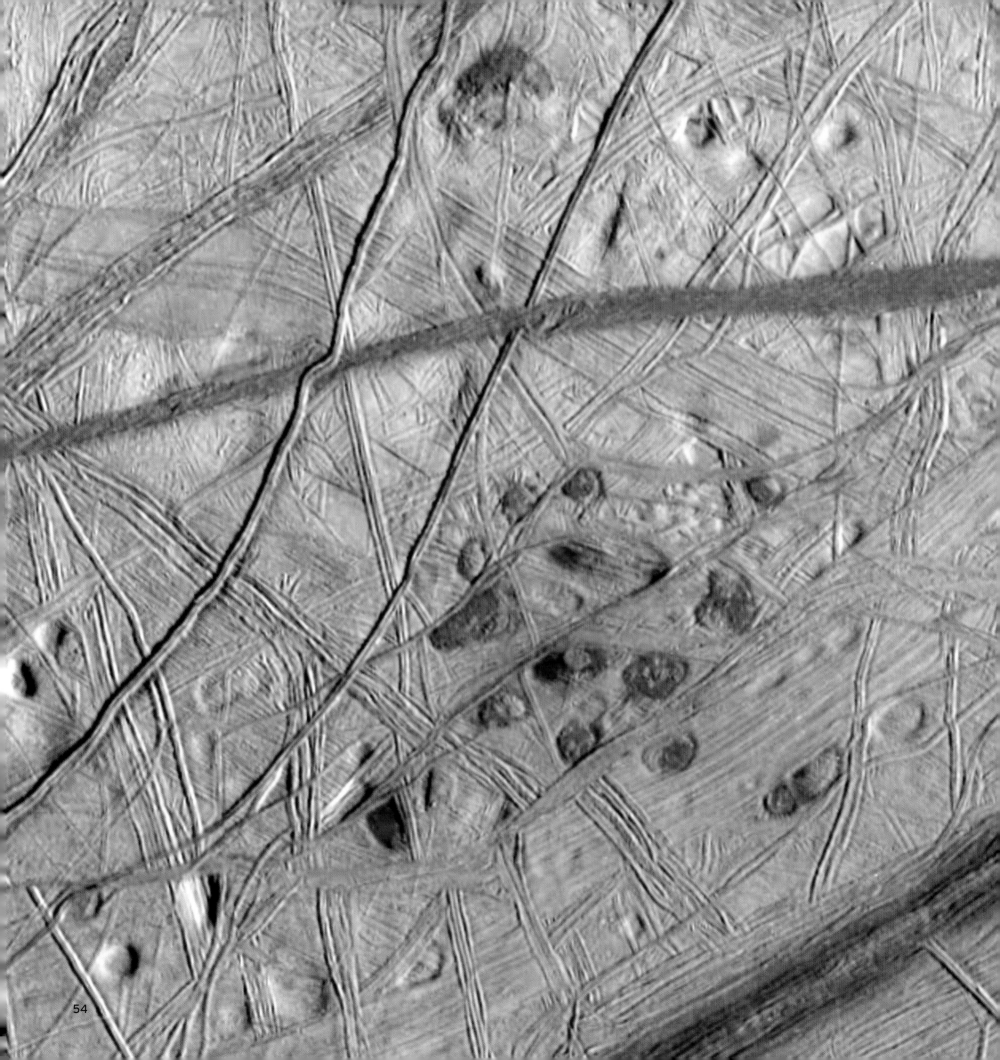

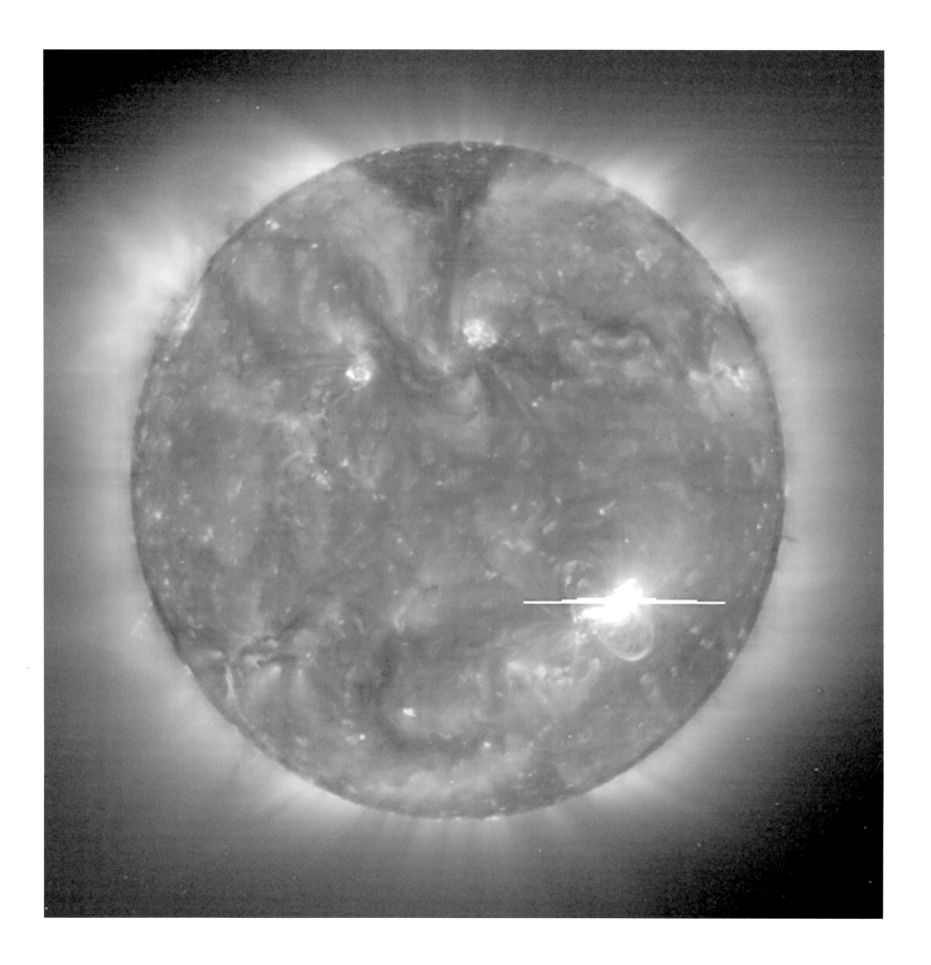

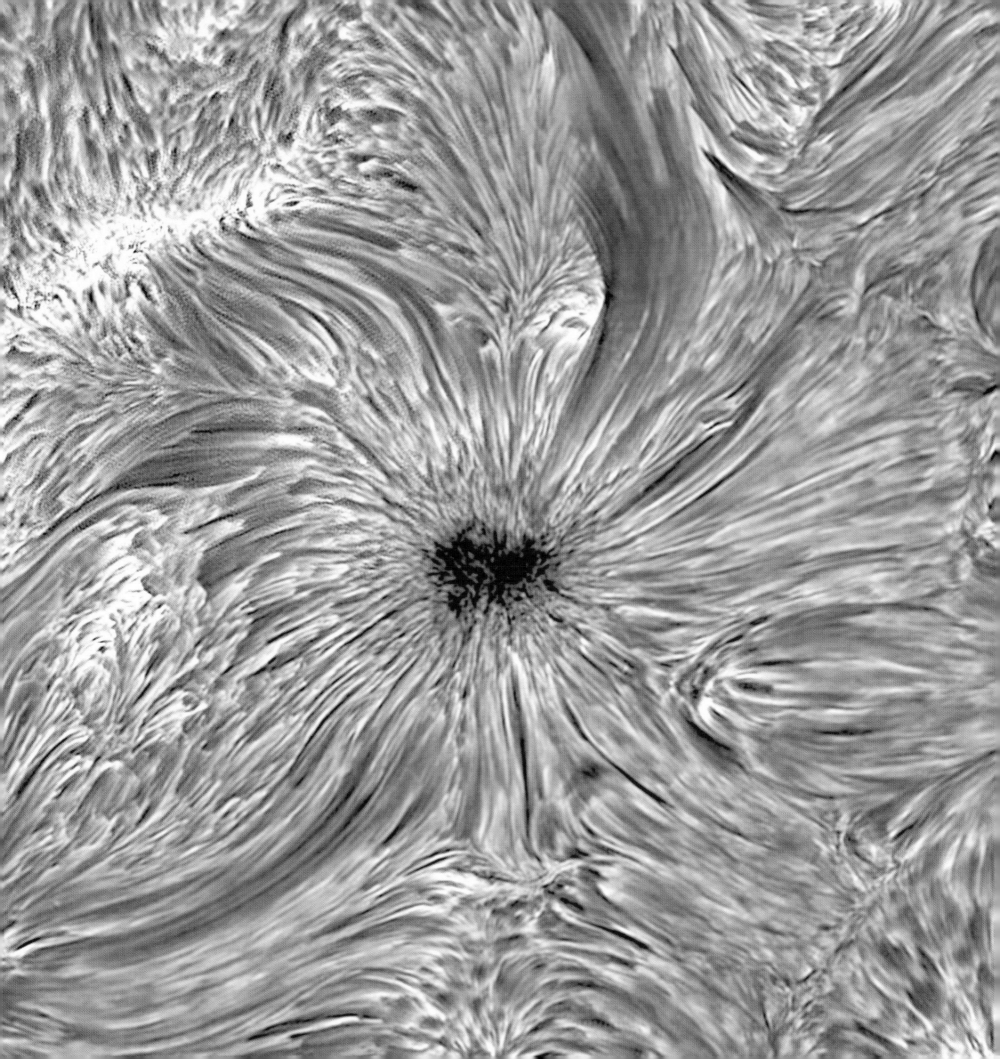

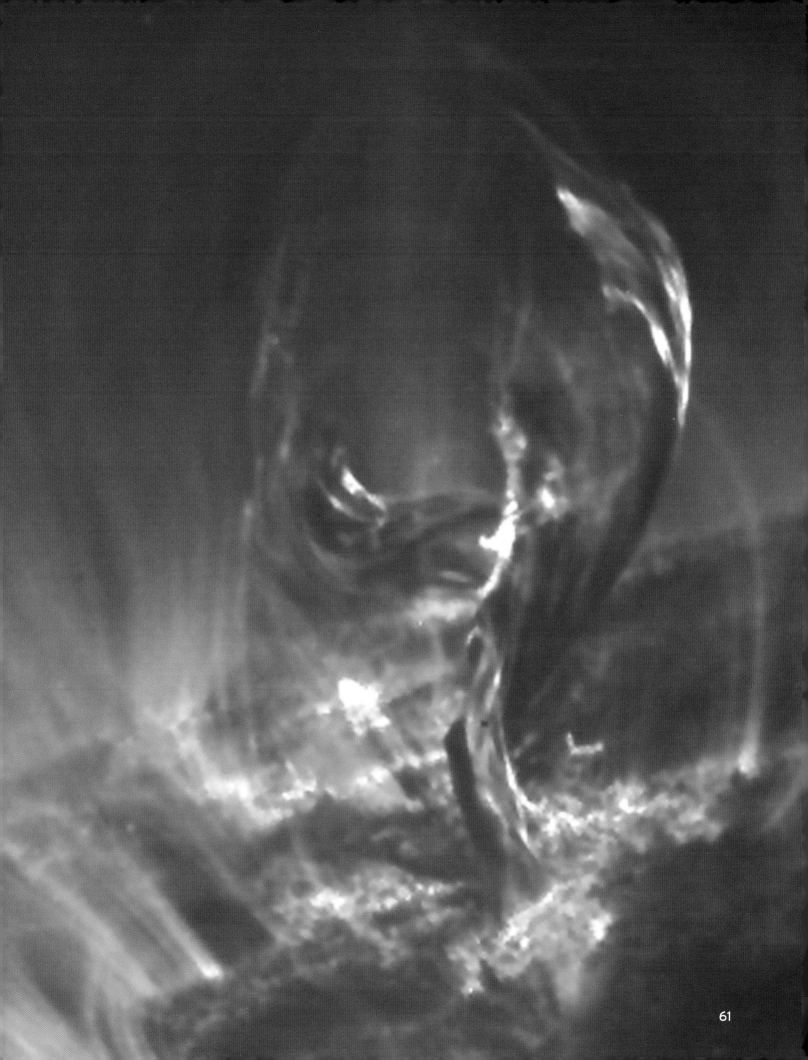

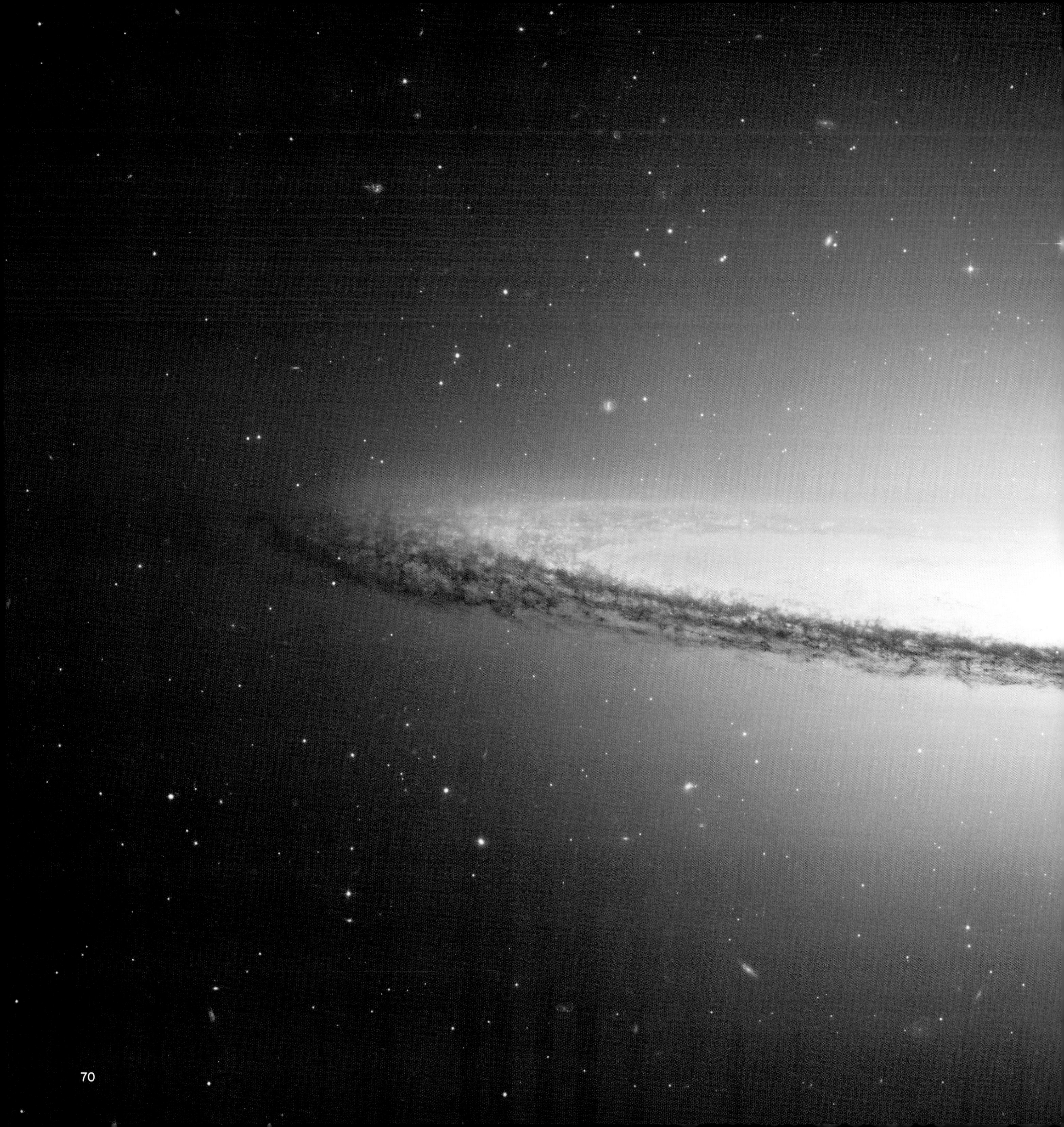

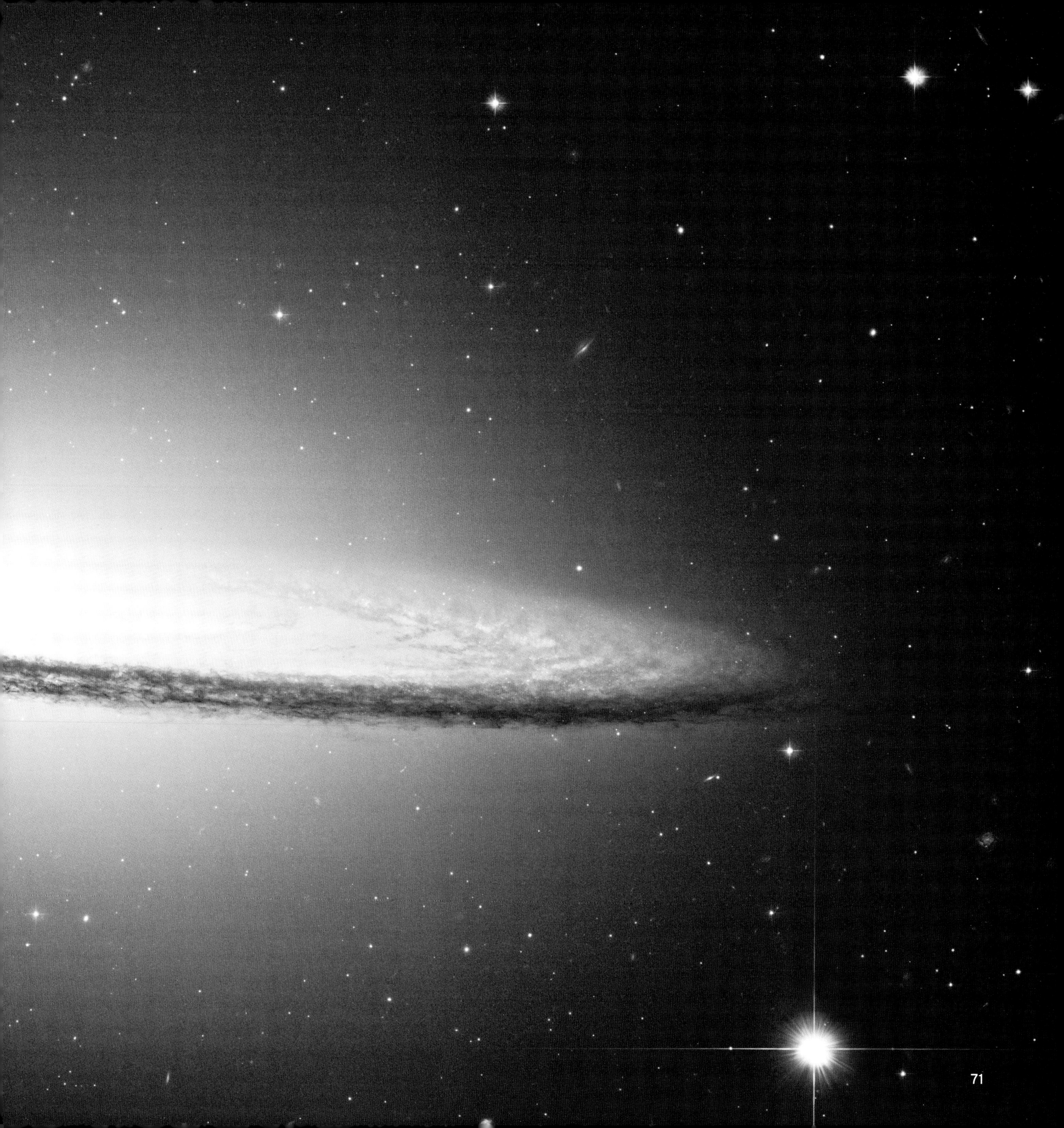

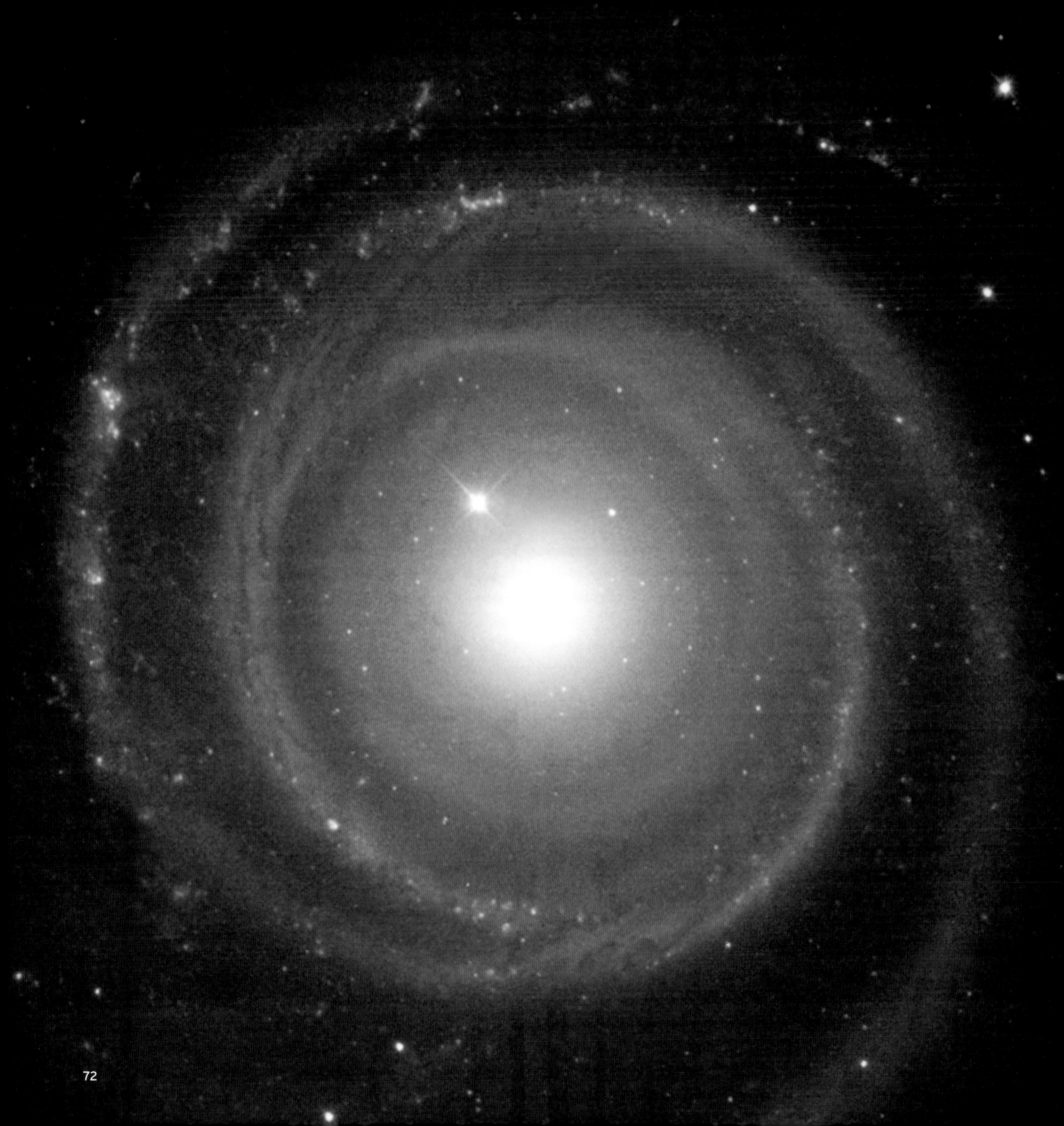

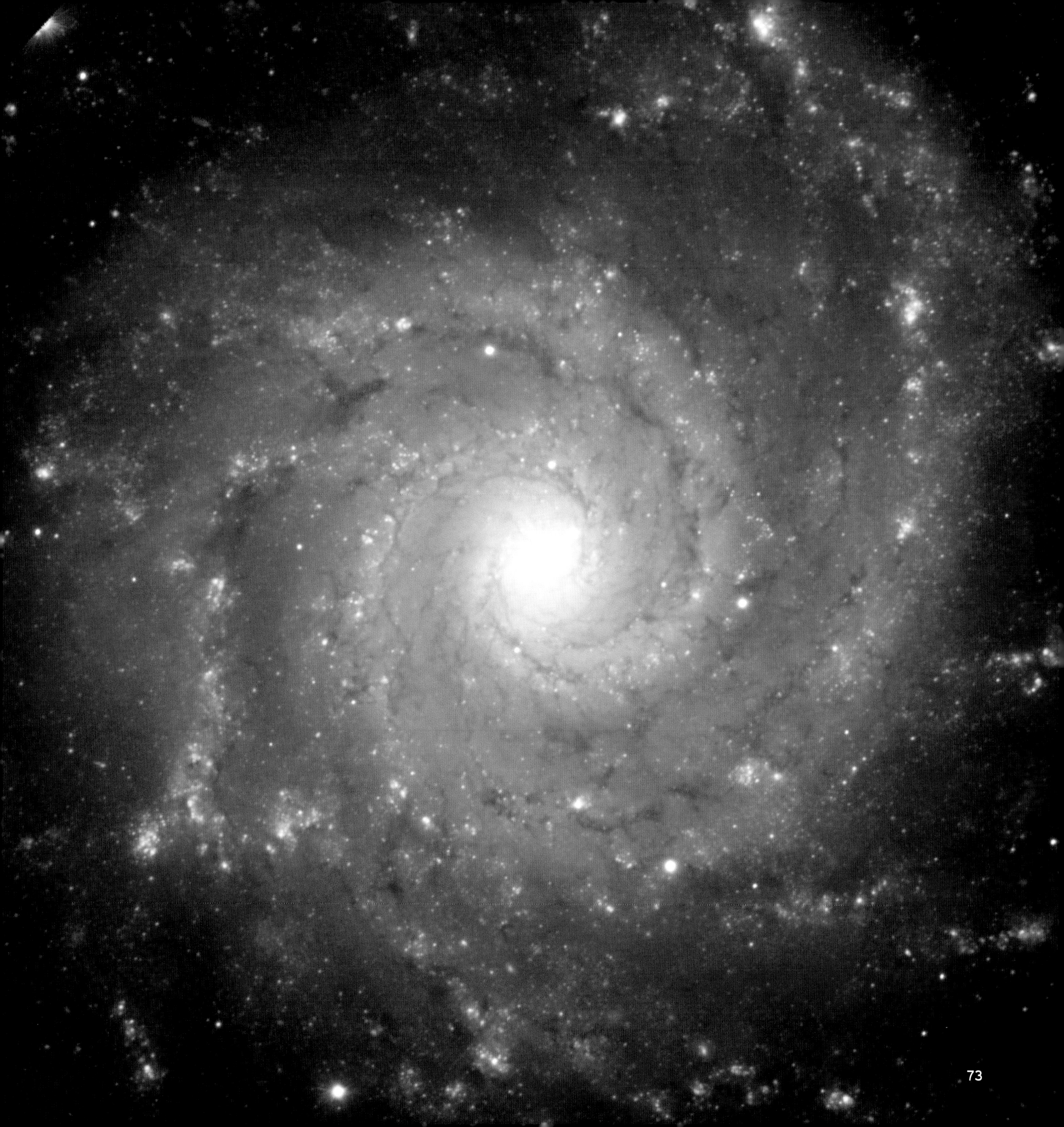

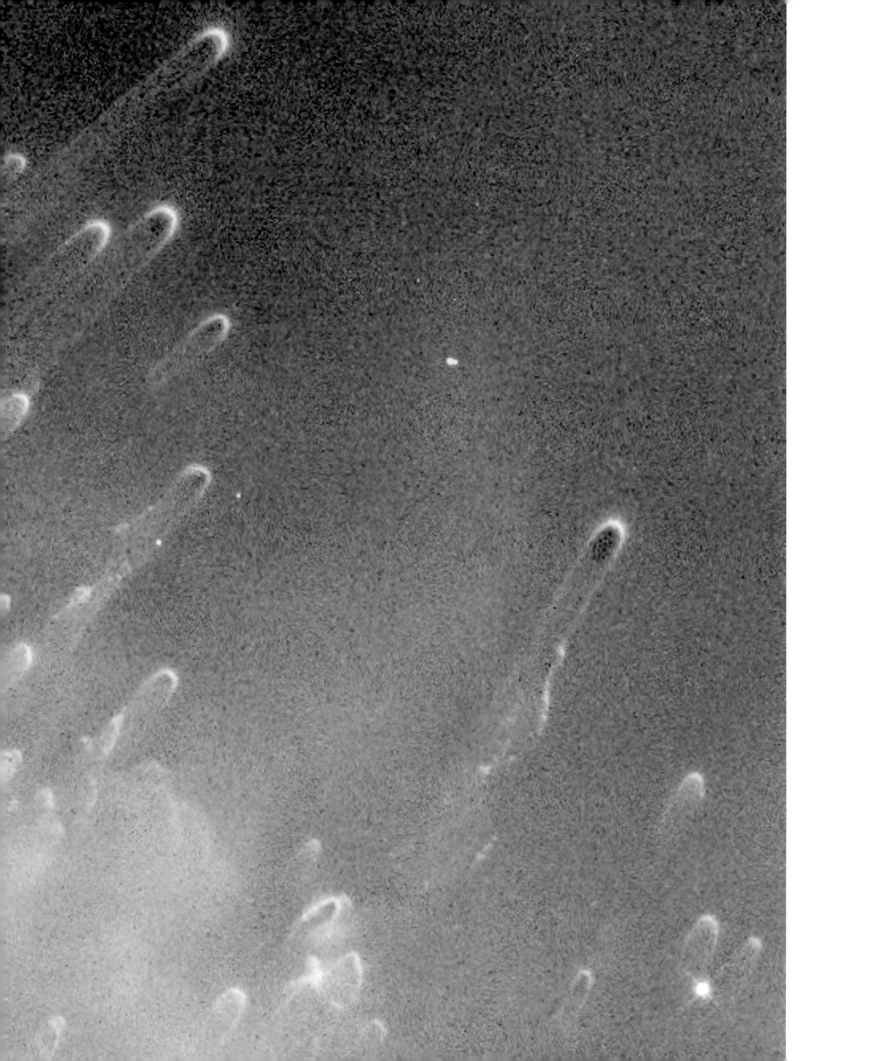

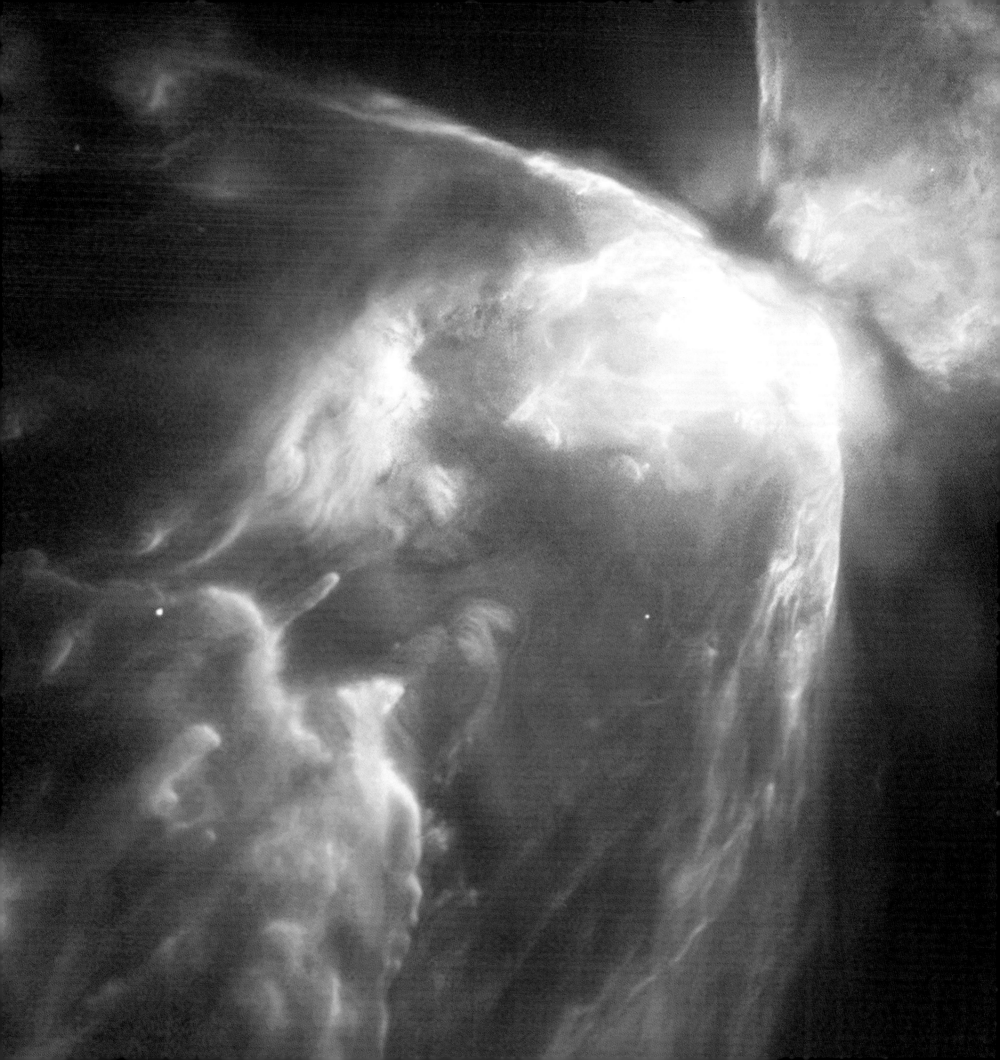

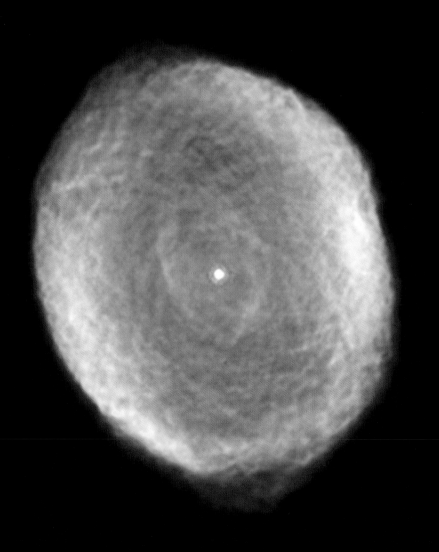

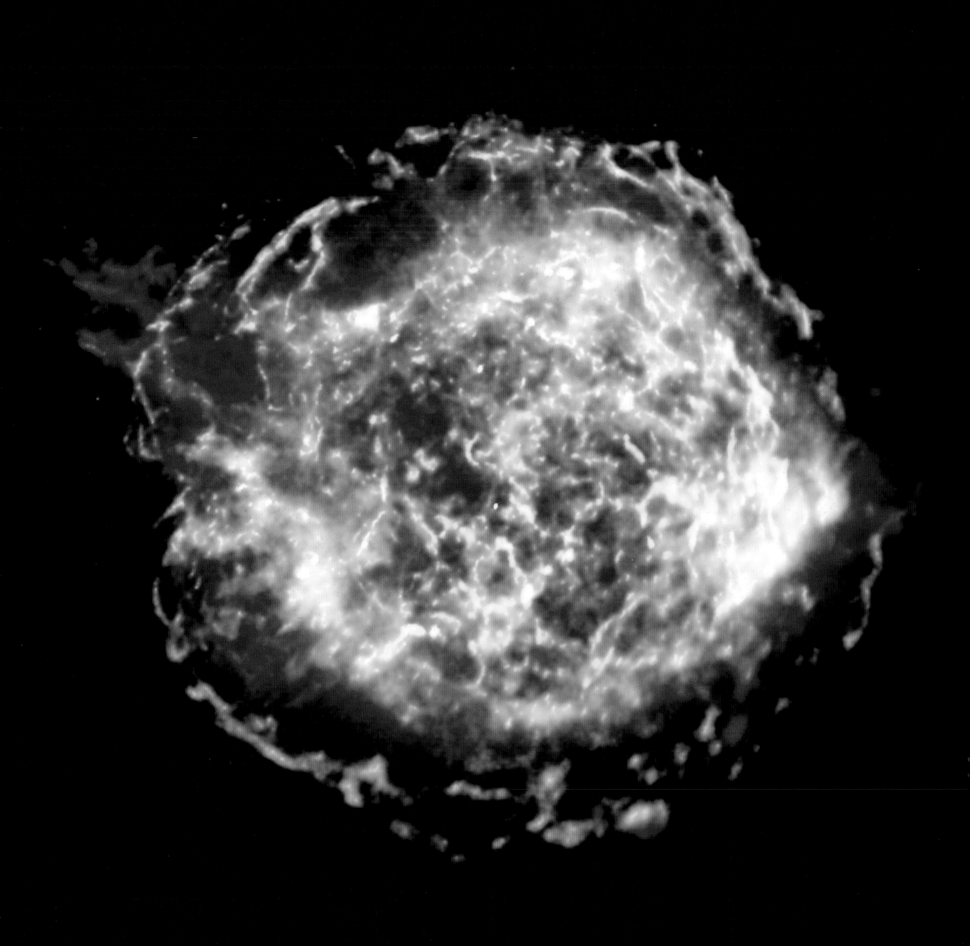

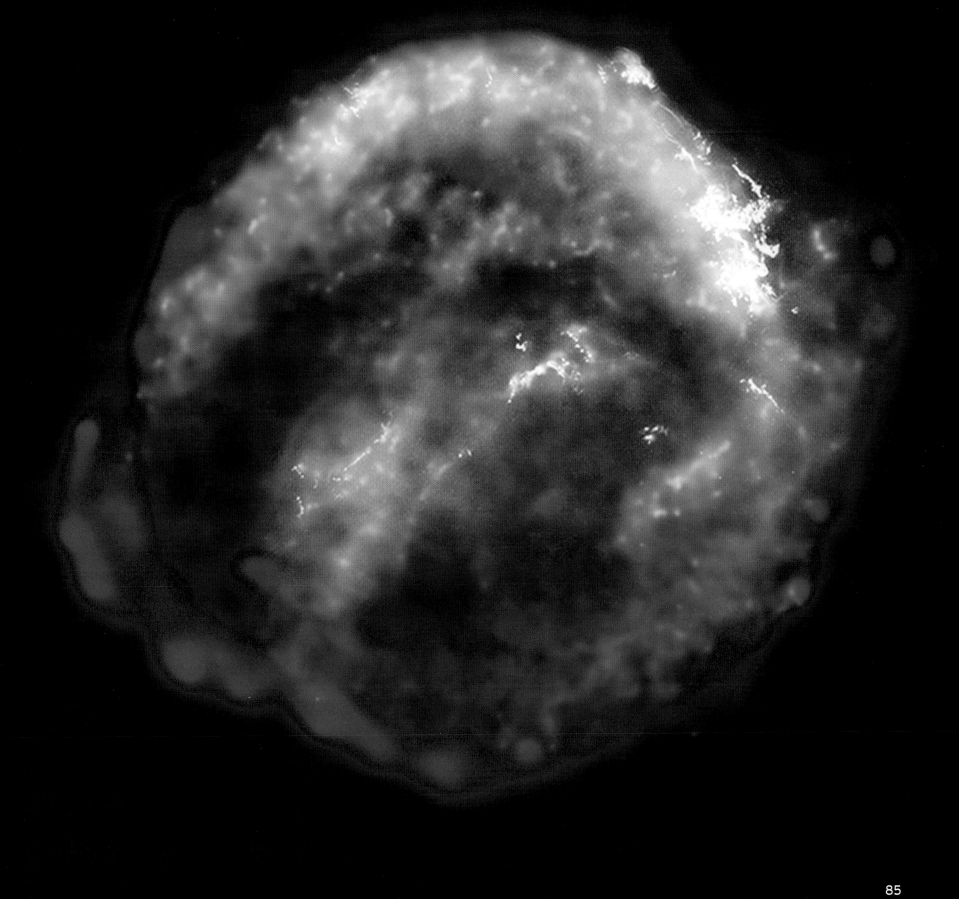

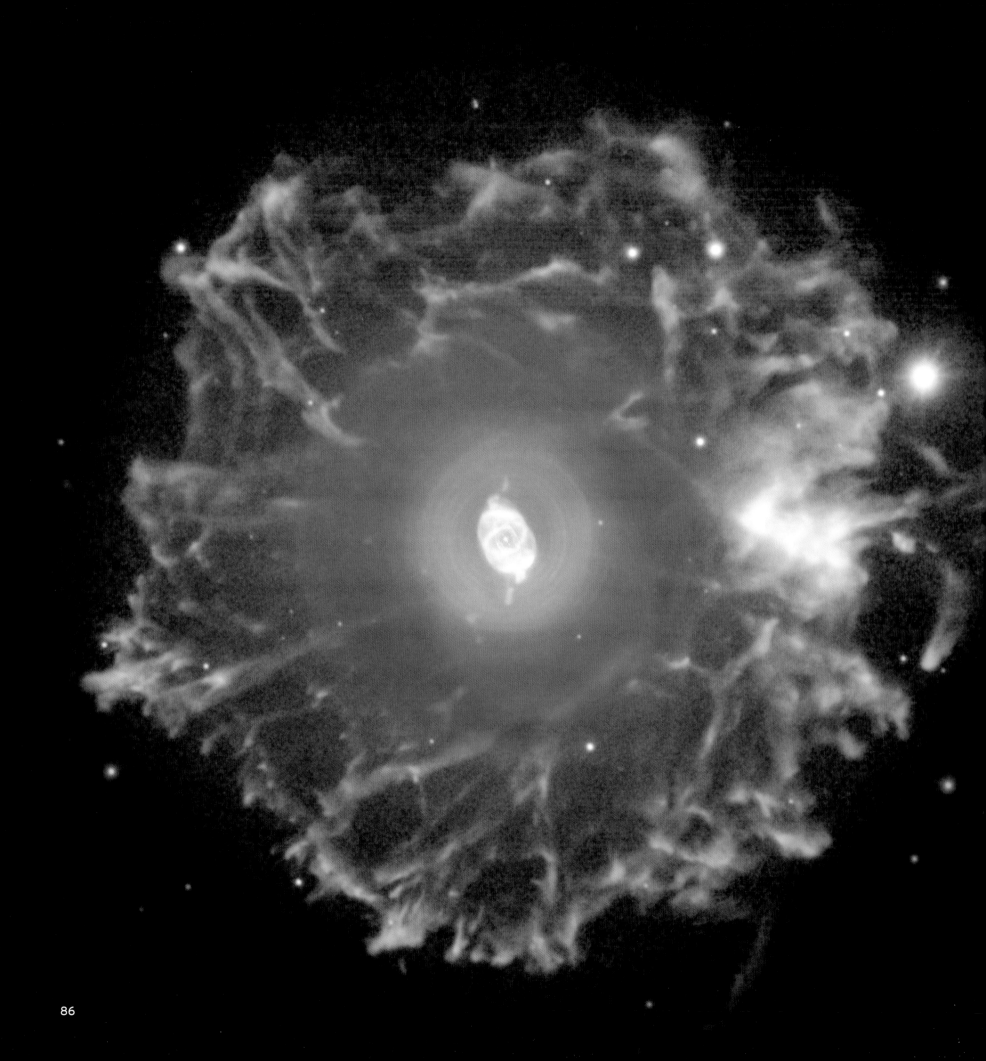

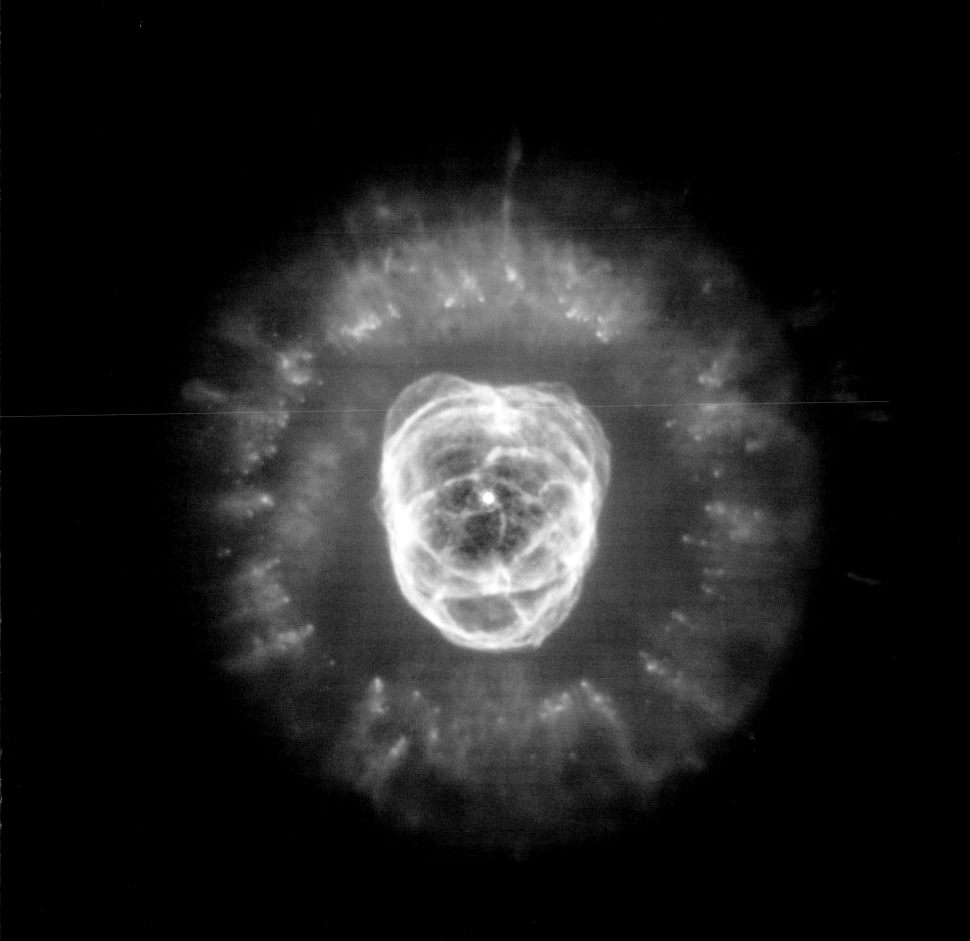

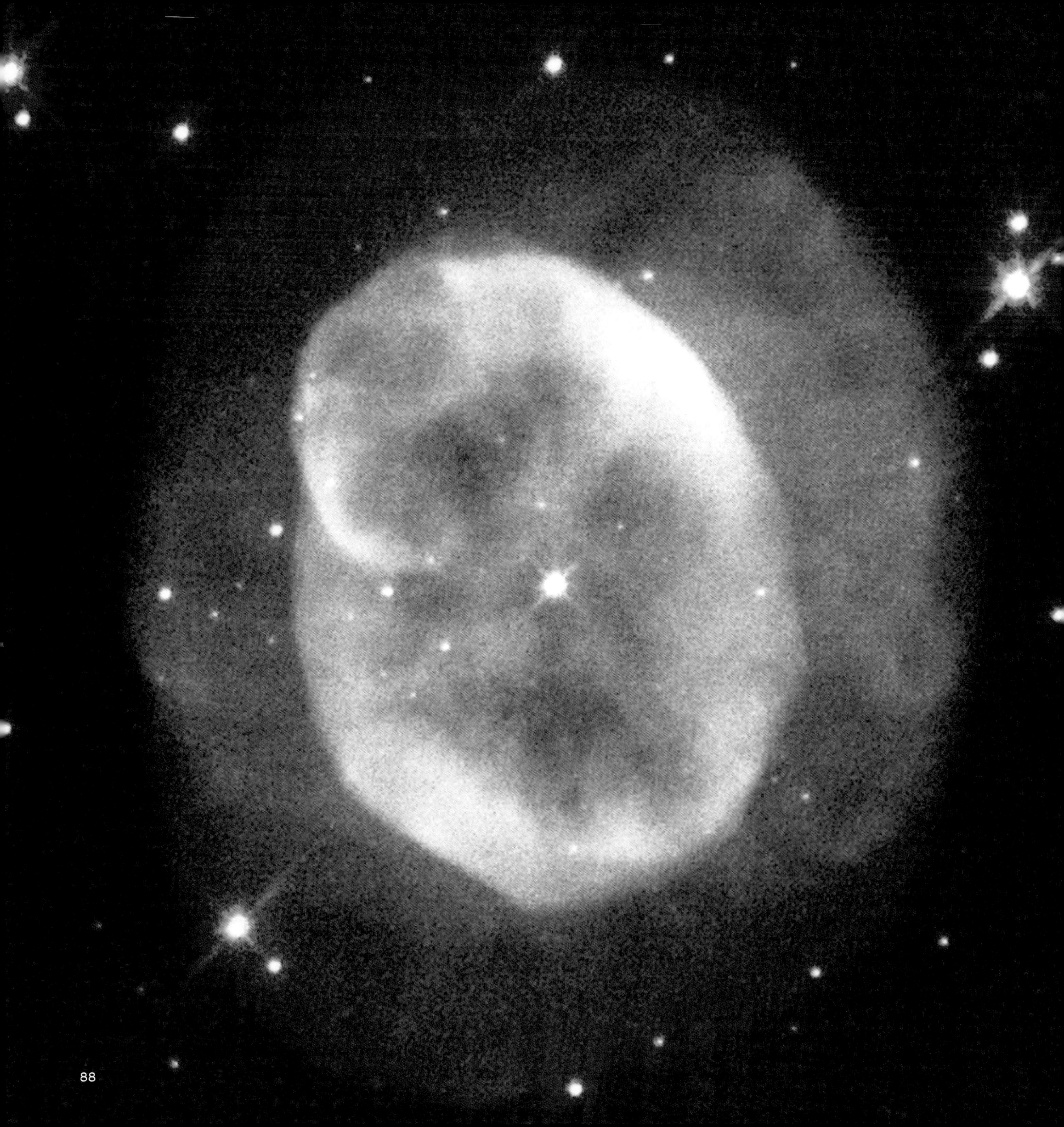

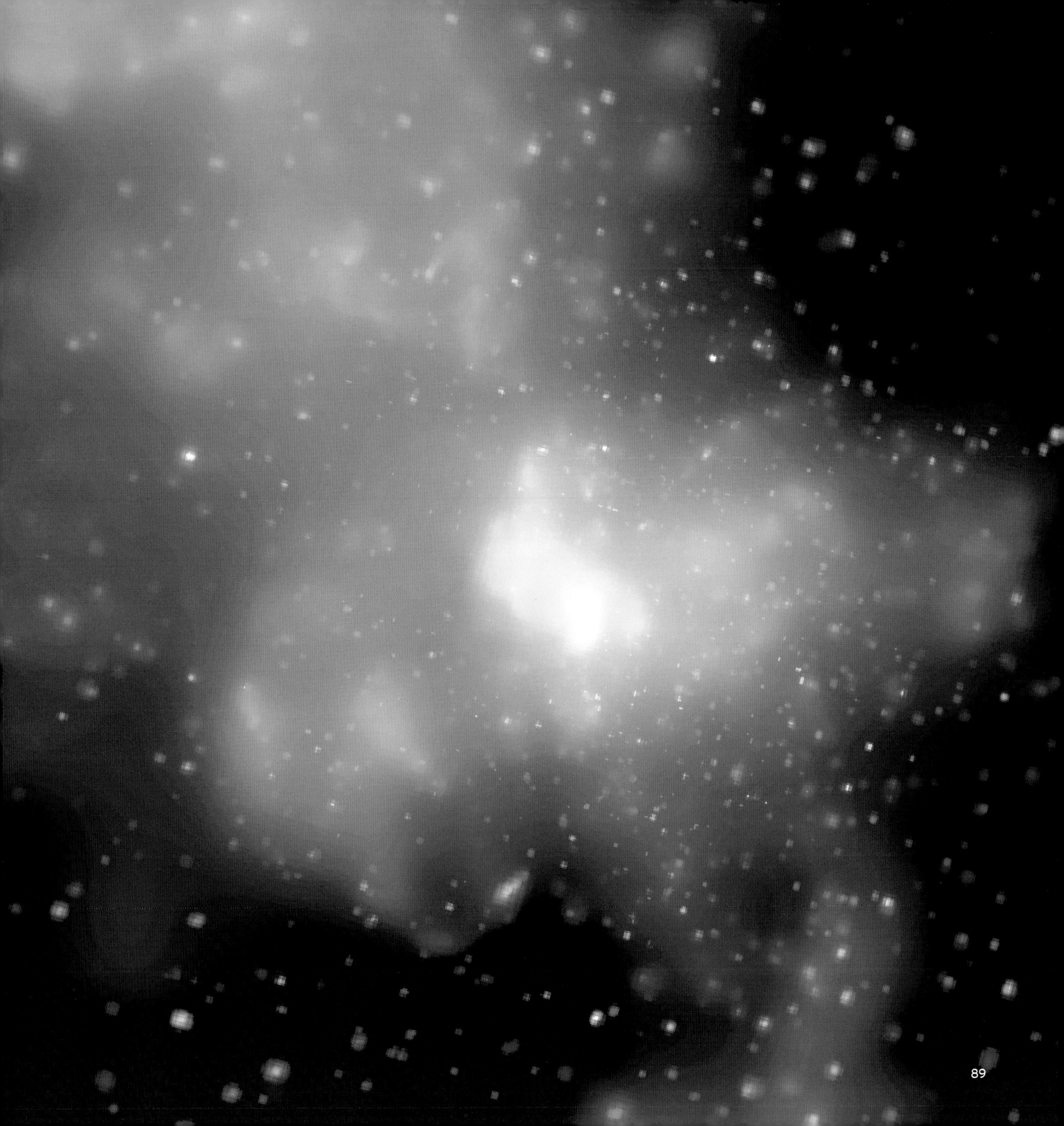

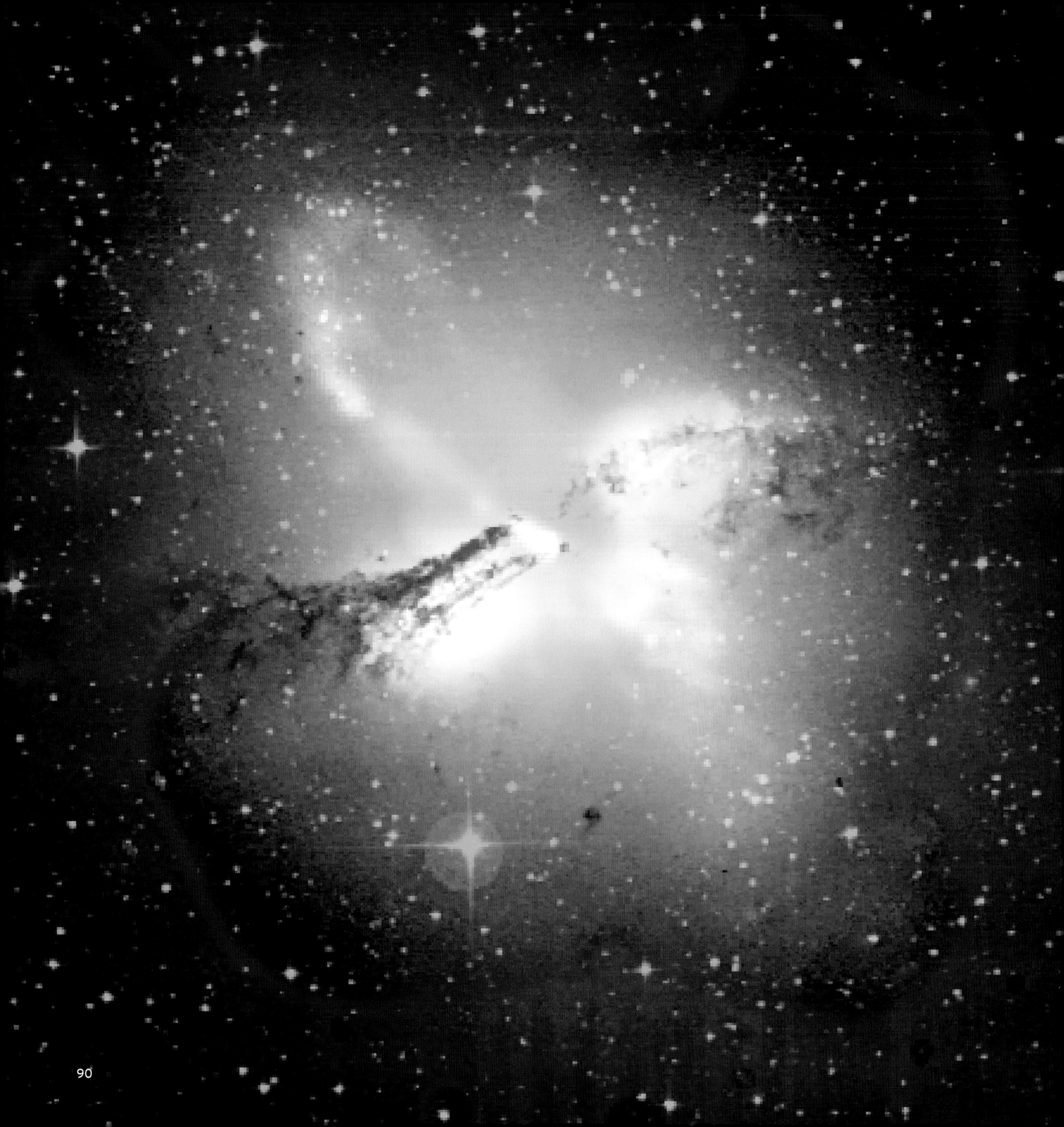

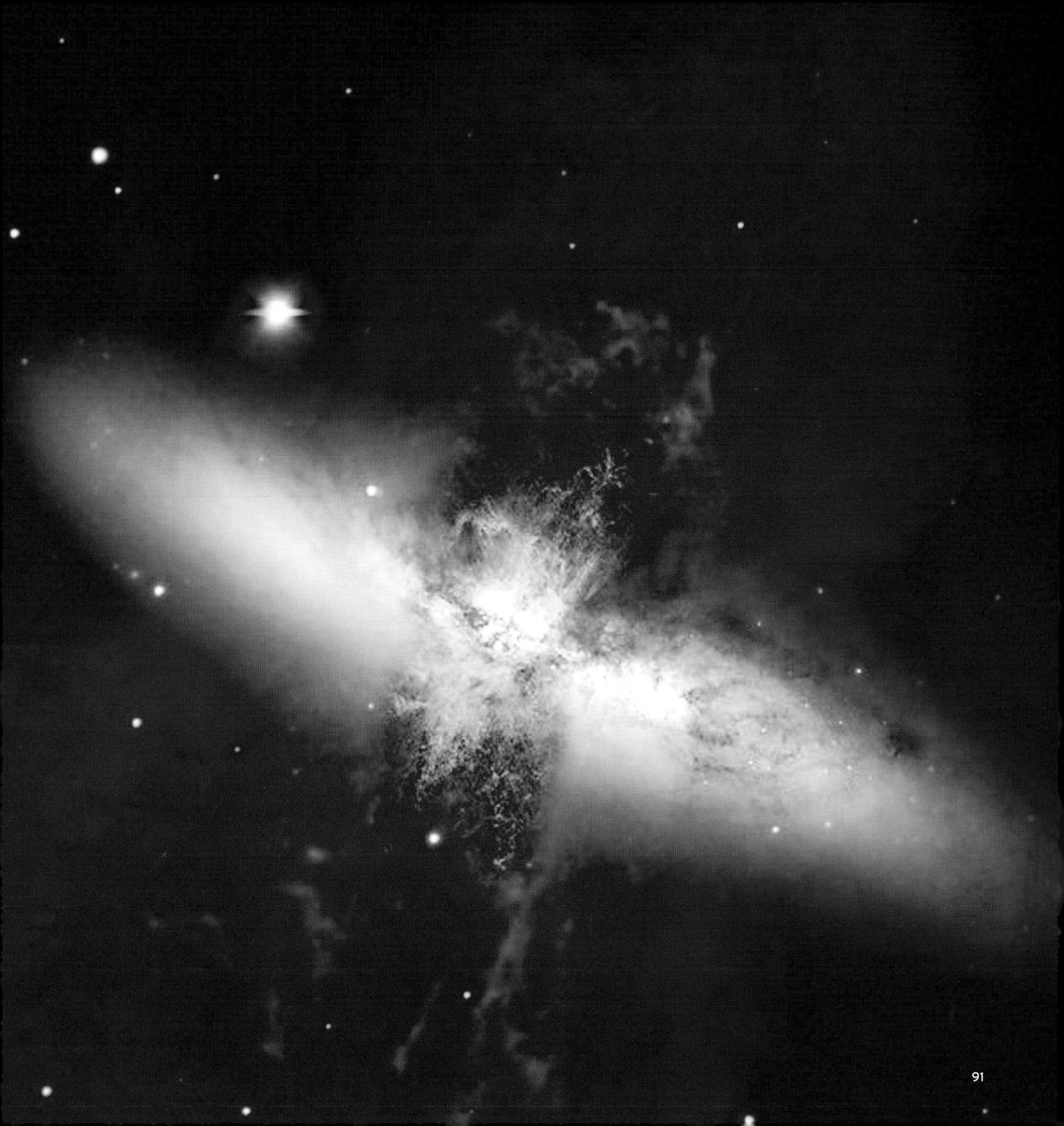

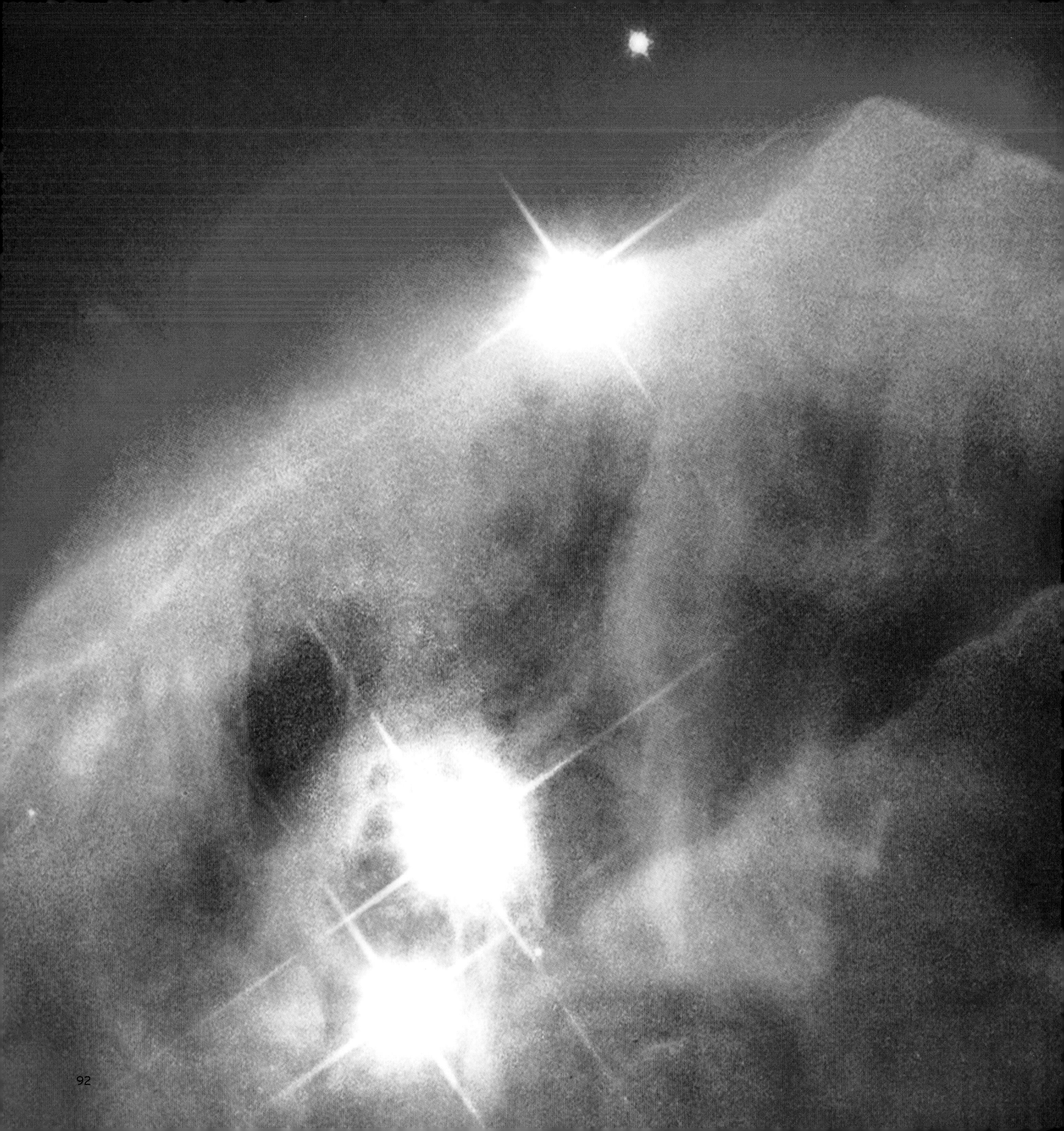

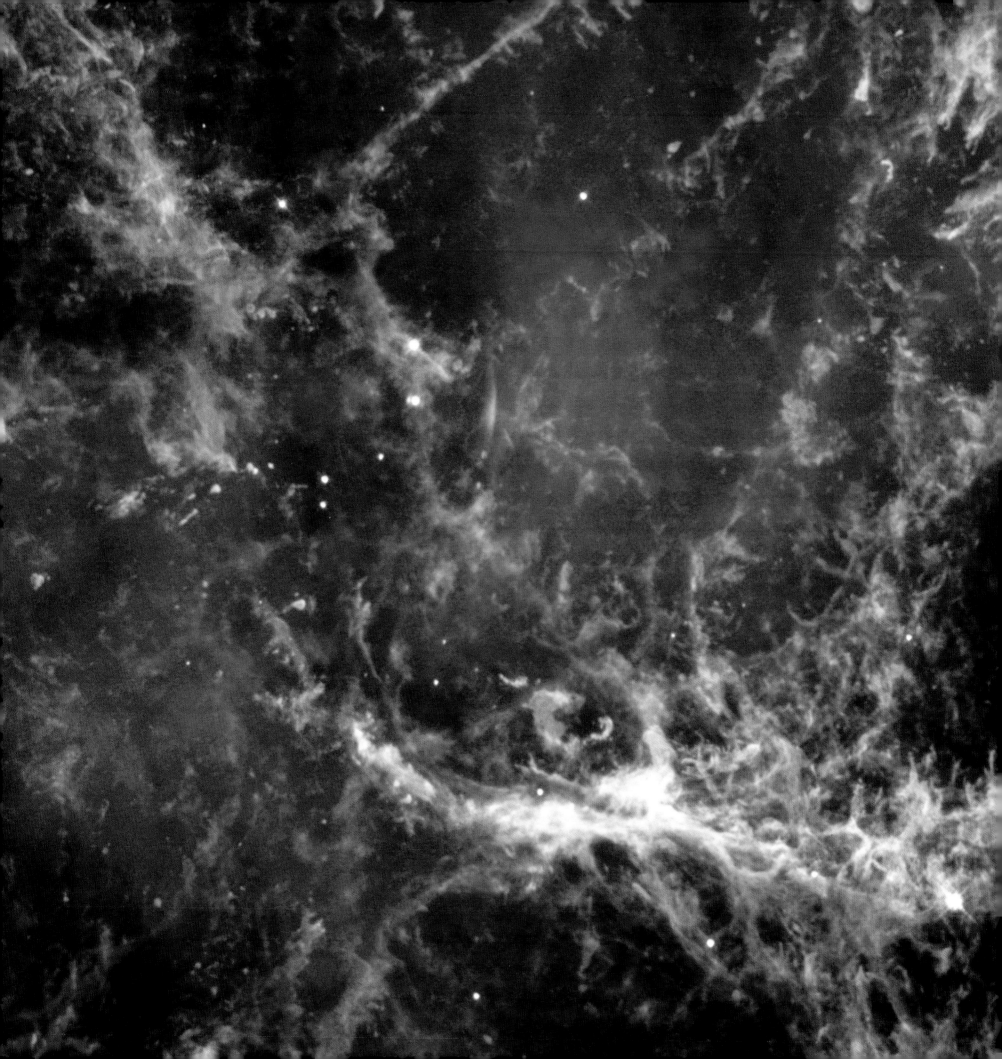

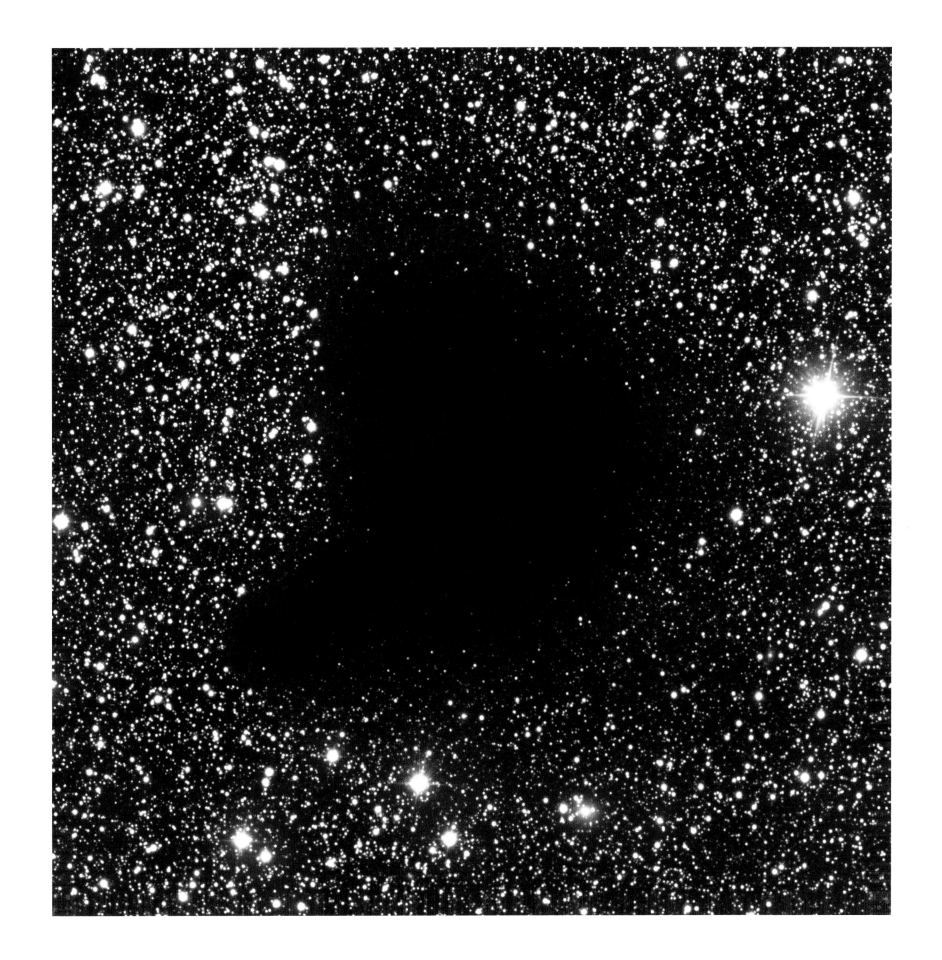

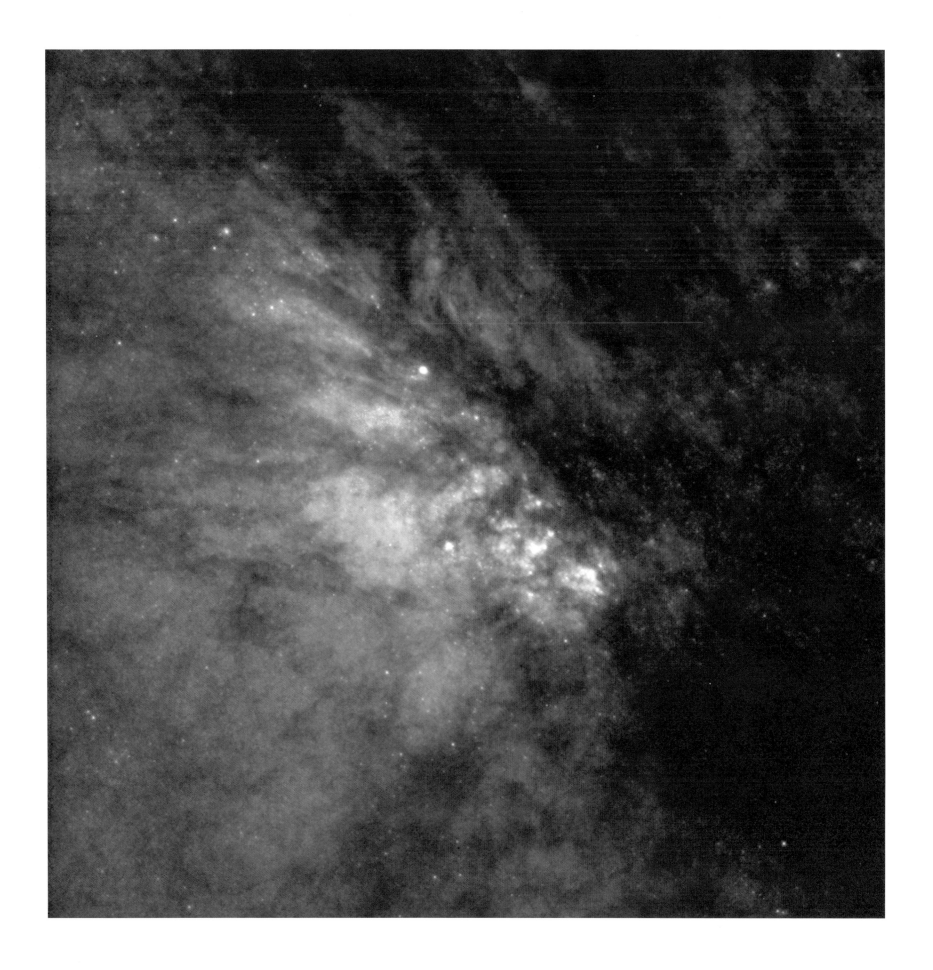

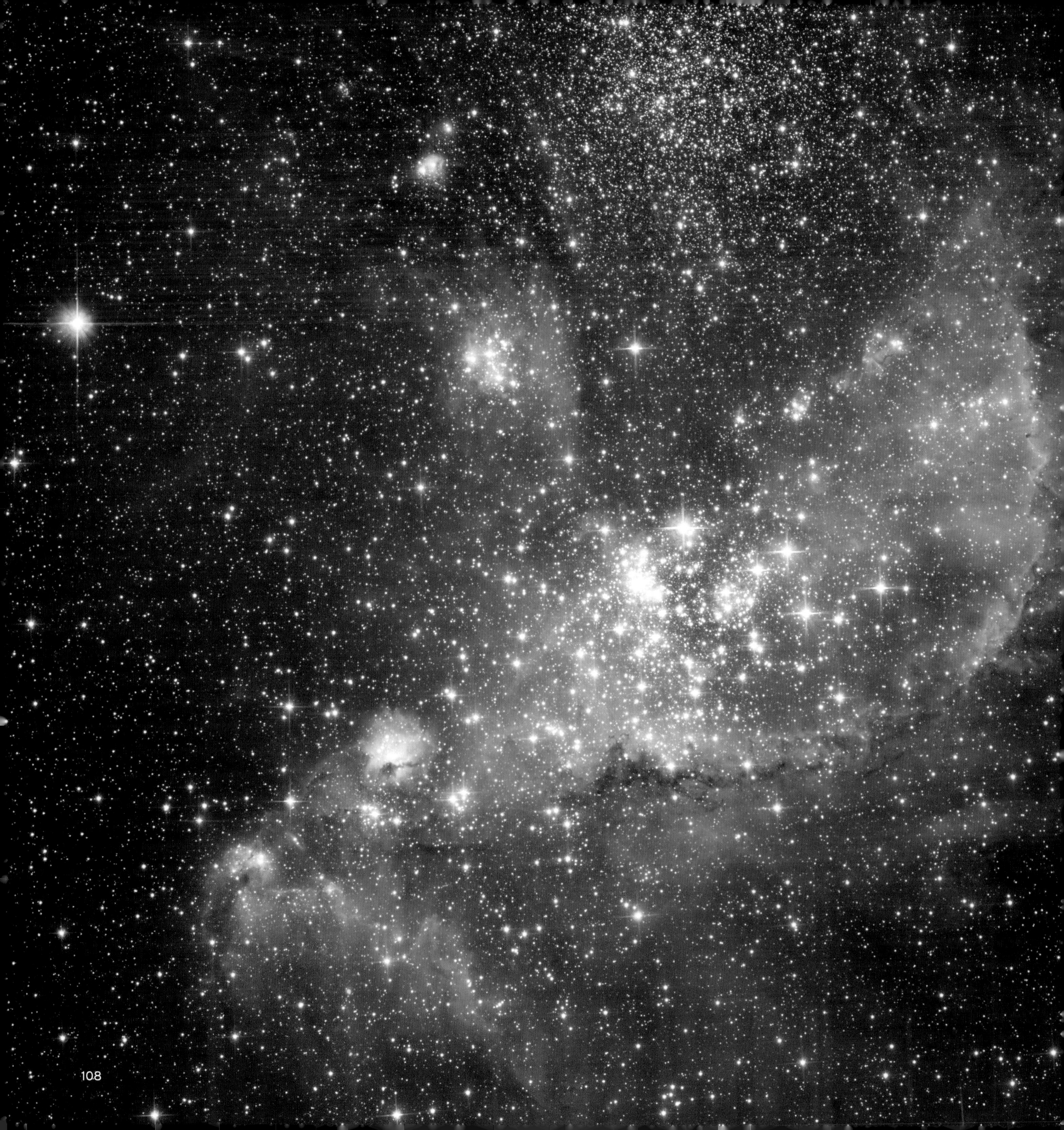

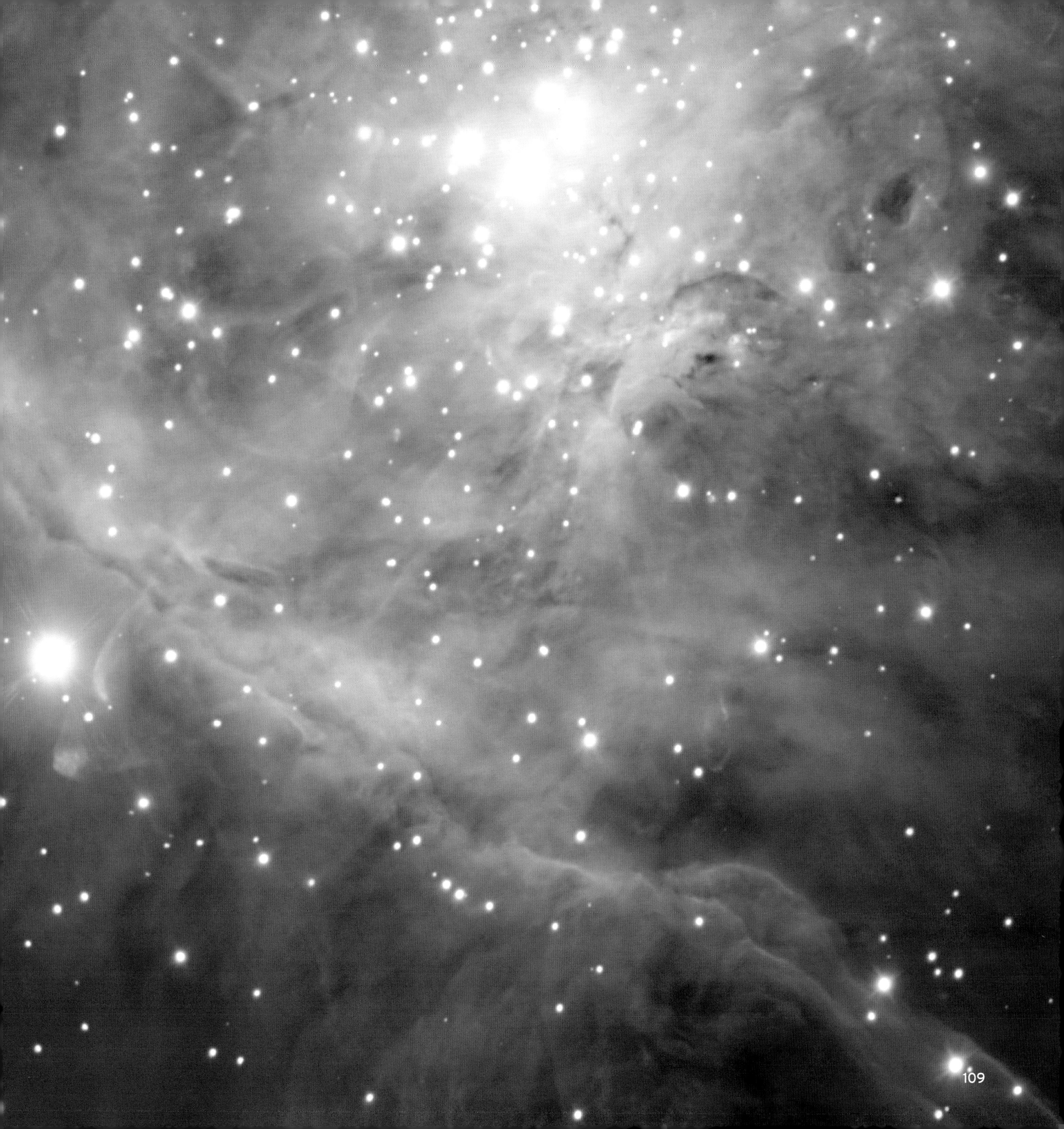

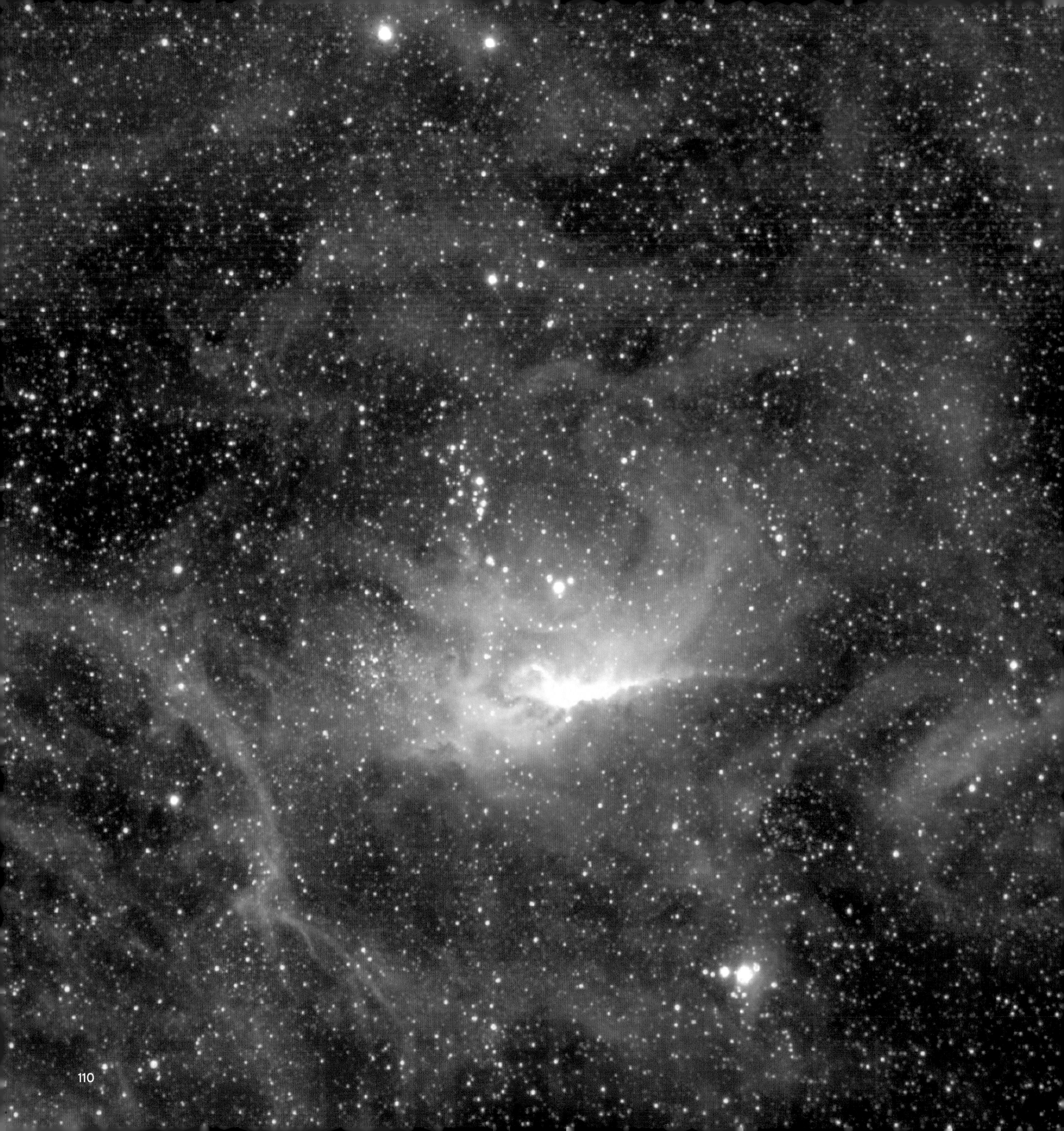

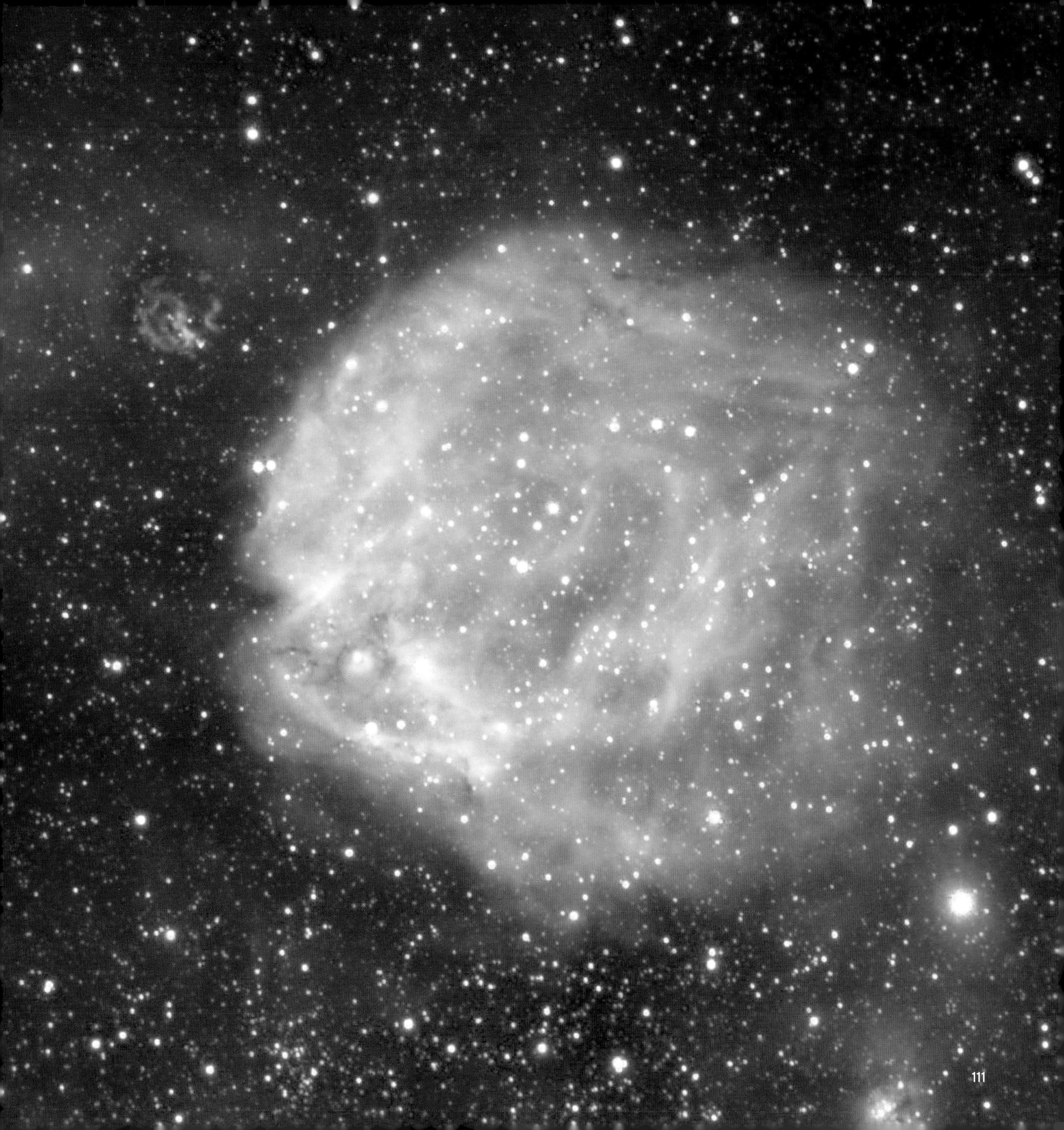

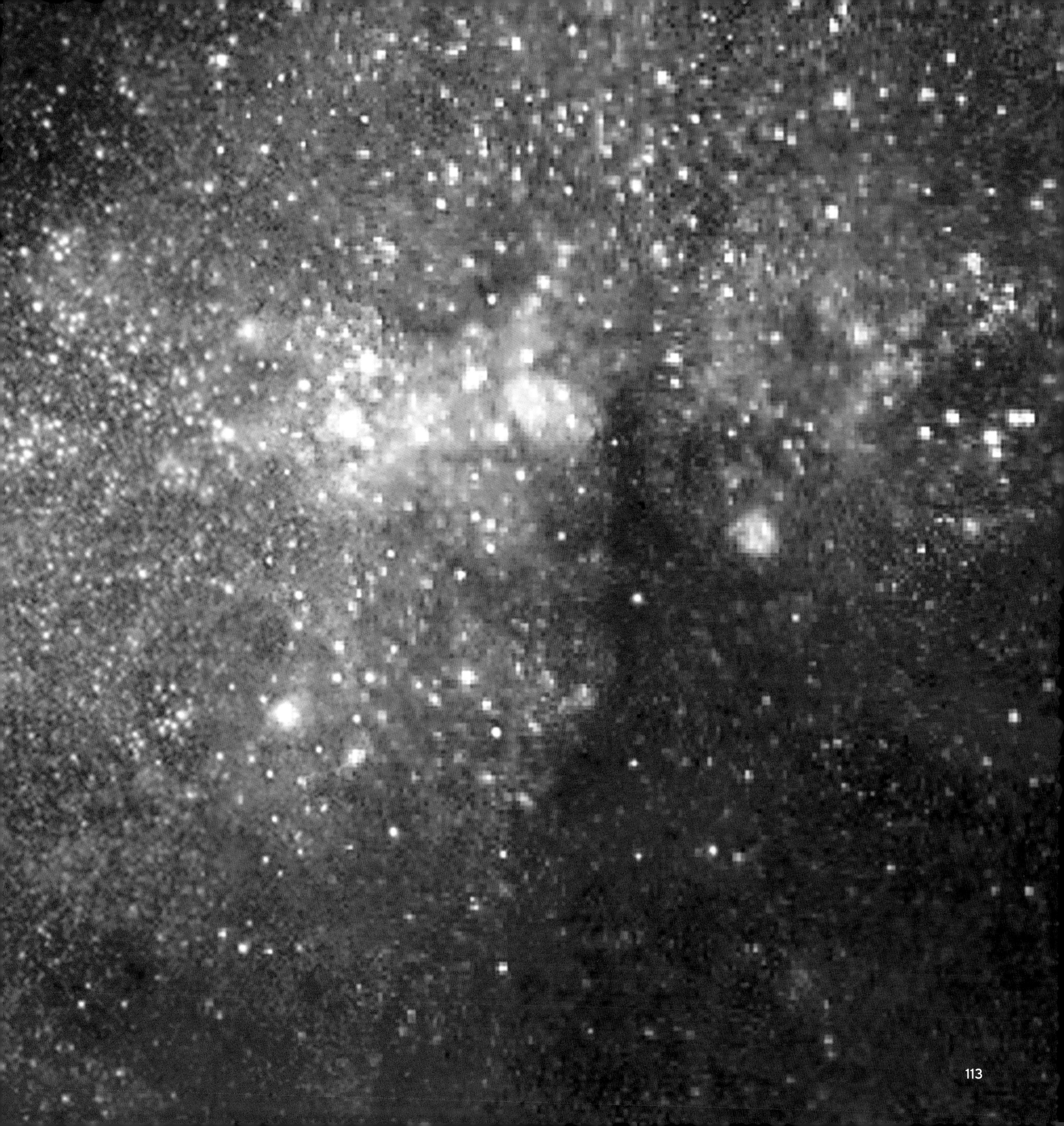

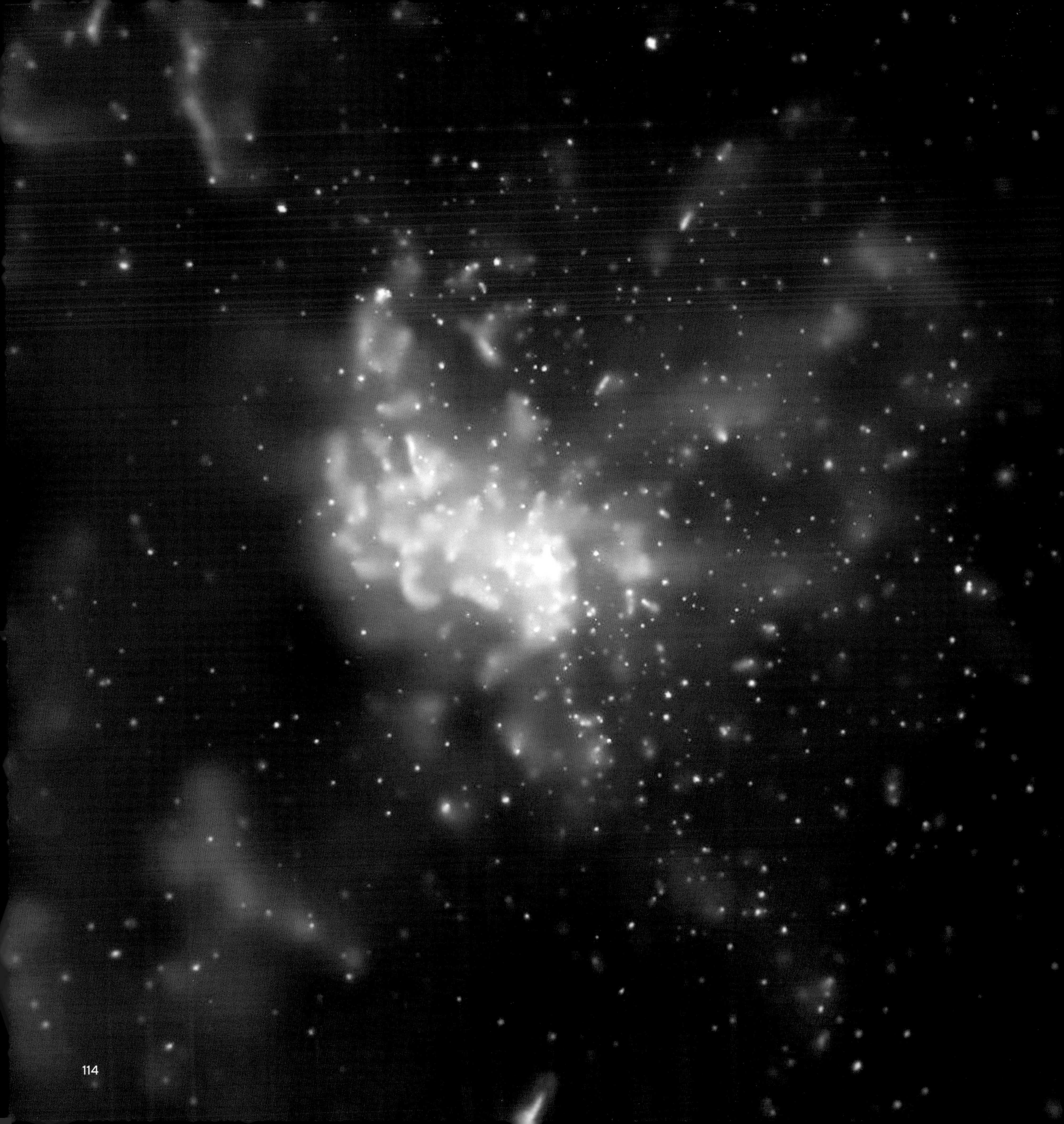

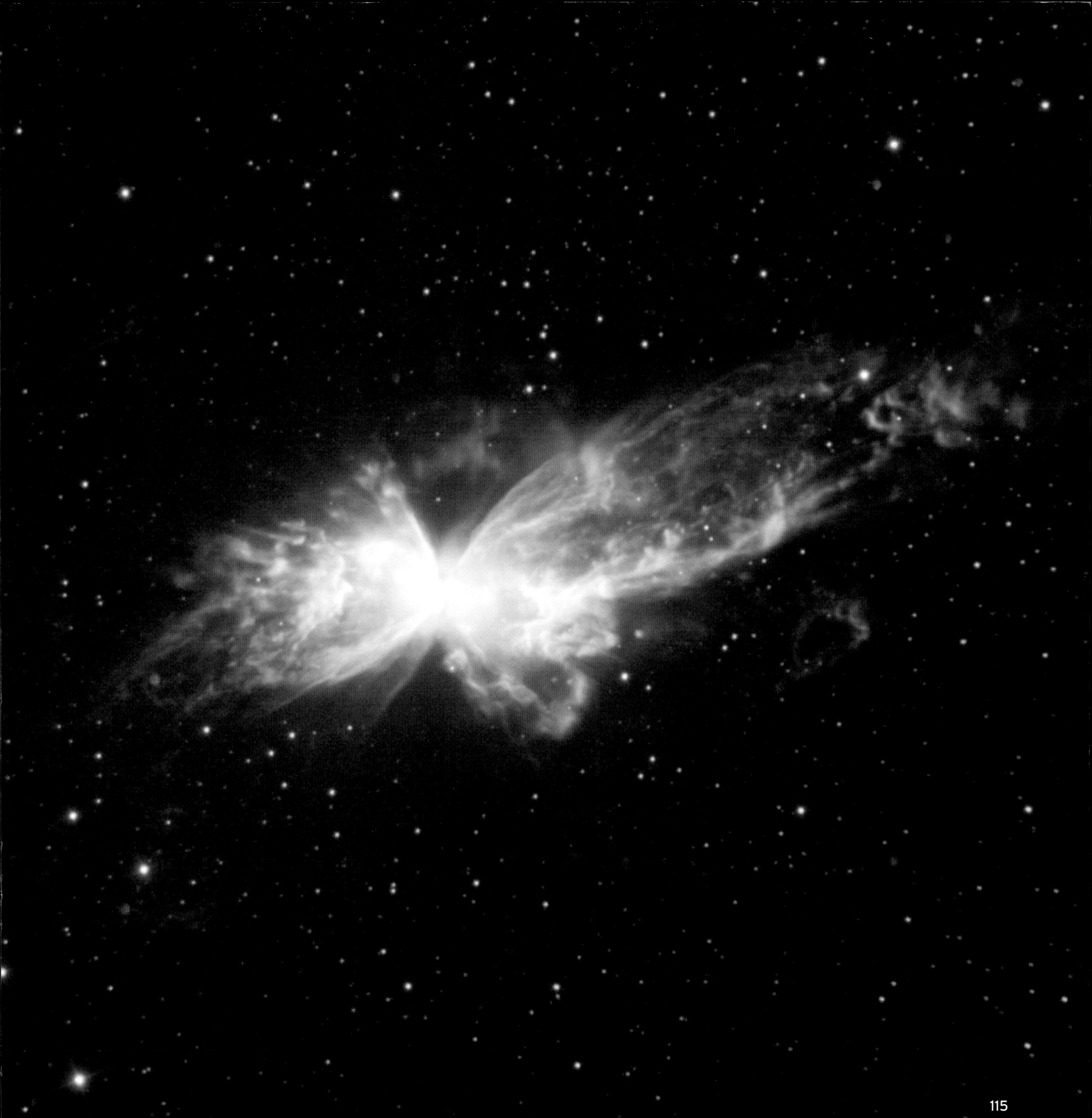

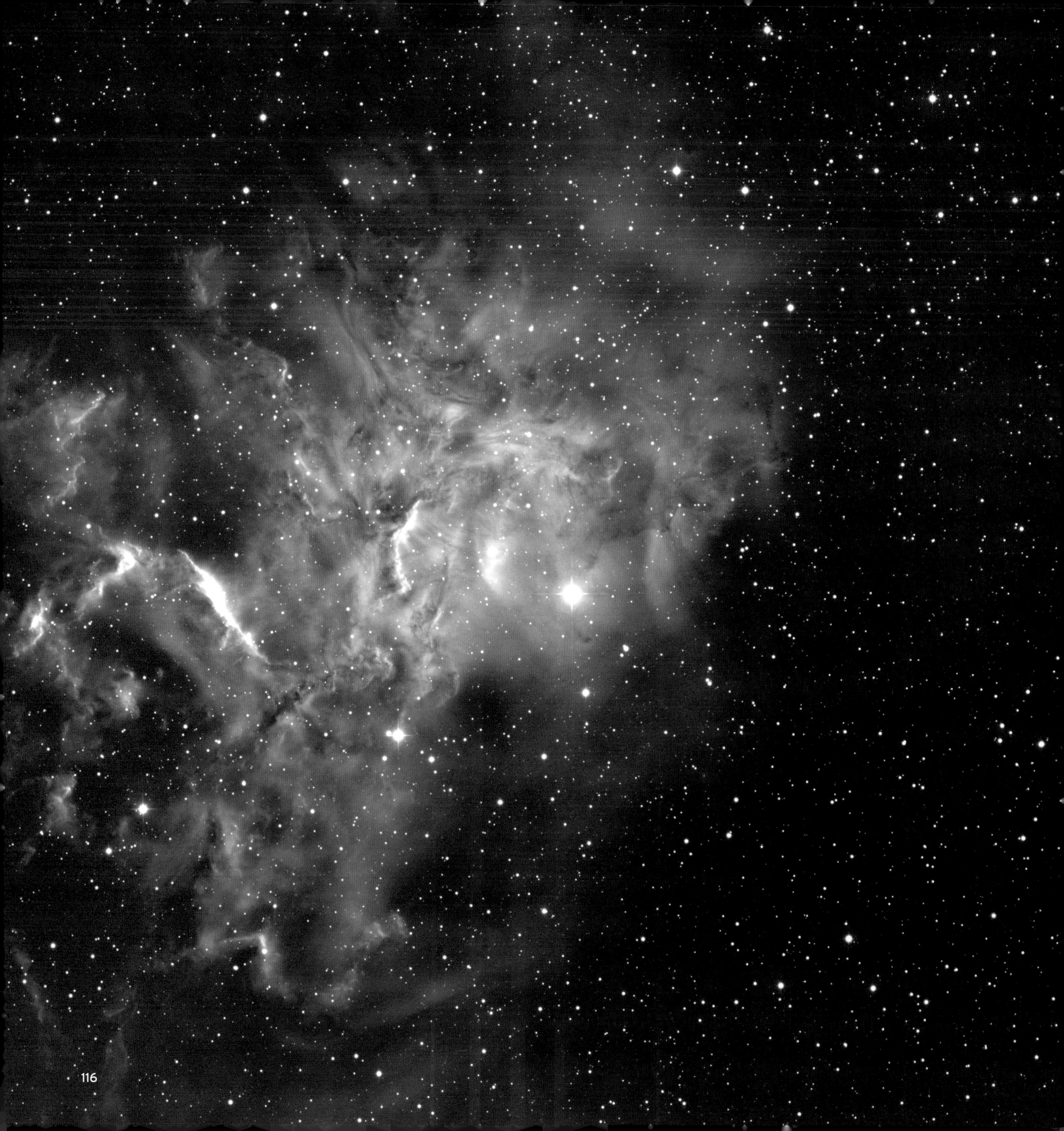

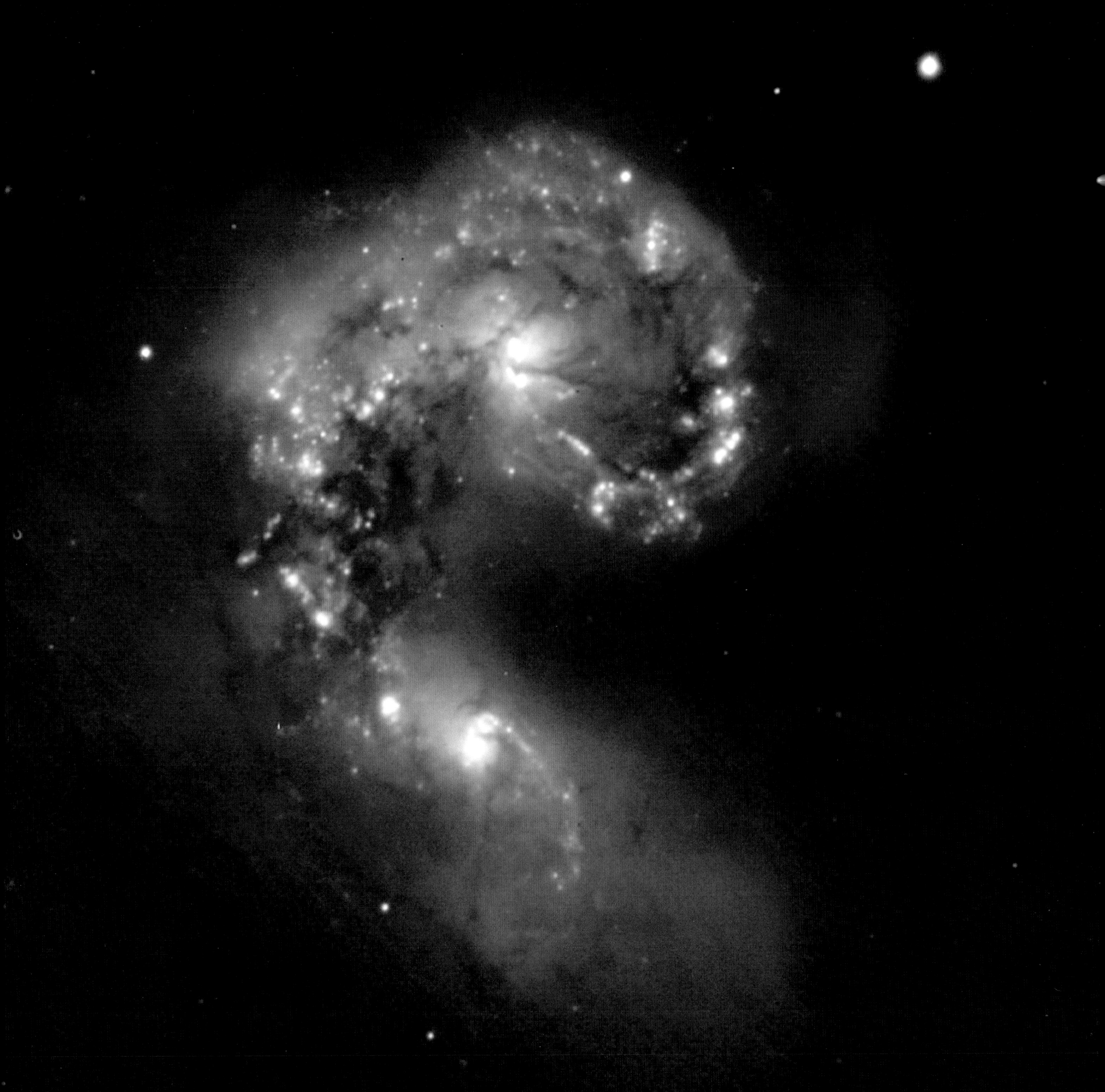

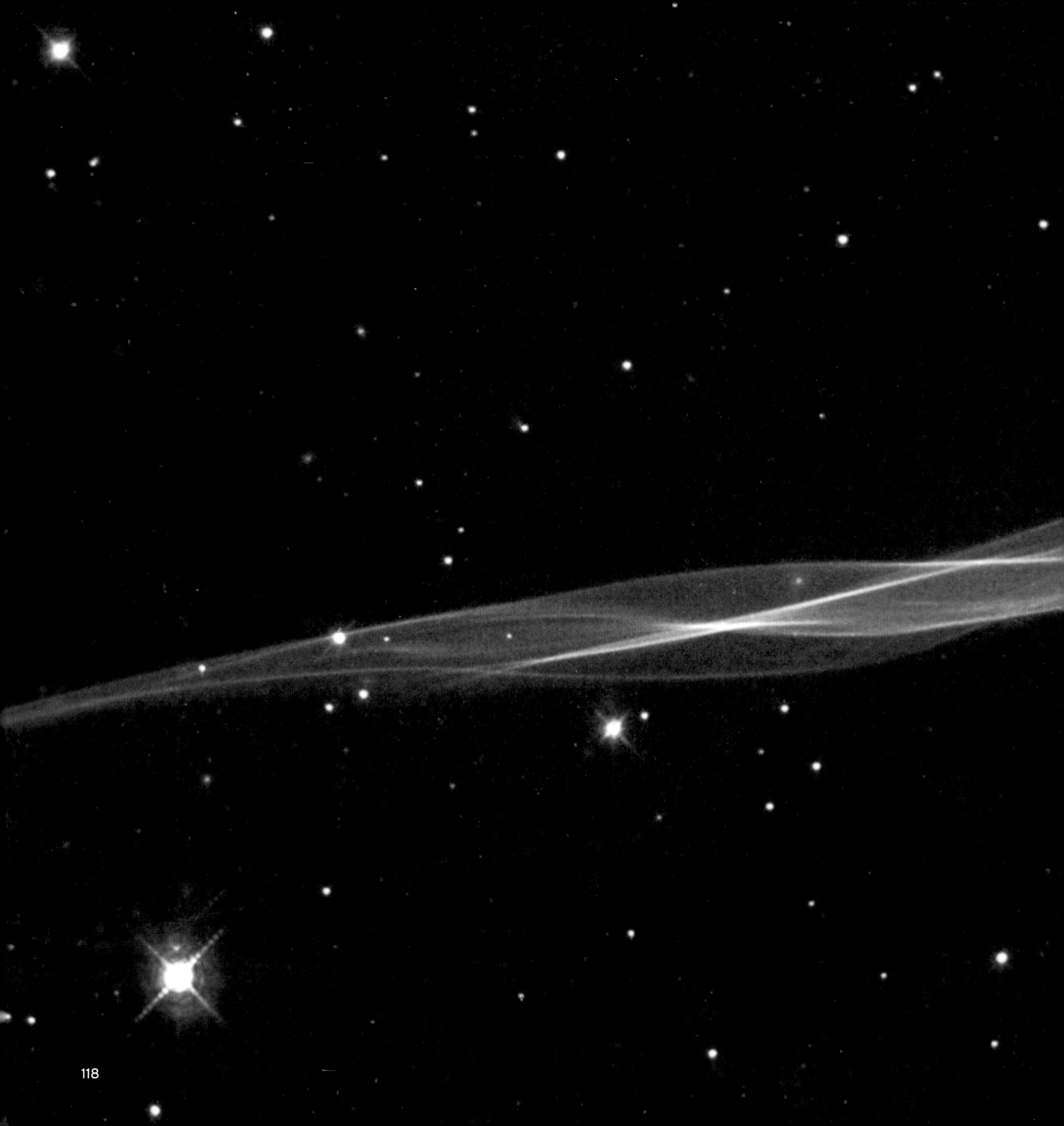

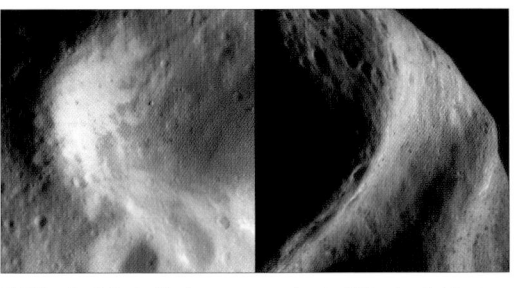

001 **Auroral oval.** The atmosphere around Jupiter's magnetic north pole glows in ultraviolet light in a process similar to the origin of the northern lights (aurora borealis) on Earth. The UV image was captured by the Hubble Space Telescope.
© *NASA/ESA/John T. Clarke (University of Michigan)*

002 **Magnificent desolation.** Apollo 17 astronaut and geologist Harrison Schmitt enjoys his stroll on the Moon in this panoramic mosaic of images made by commander Eugene Cernan. To date, Schmitt and Cernan are the last human beings to visit another world.
© *NASA/Mark Spoelstra (Artis Planetarium)*

004 **Psyche delicate.** The largest crater on the asteroid Eros is called Psyche. In this closeup by the spacecraft NEAR-Shoemaker, delicate patterns of bright and dark material are seen on the crater slopes.
© *NASA/JPL/JHU-APL*

005 **Rocky saddle.** Nicknamed the Saddle, this region of Eros is the largest depression on the elongated asteroid. The photograph was taken by NEAR-Shoemaker, which orbited the space rock at an altitude of just 100 kilometres in 2000.
© *NASA/JPL/JHU-APL*

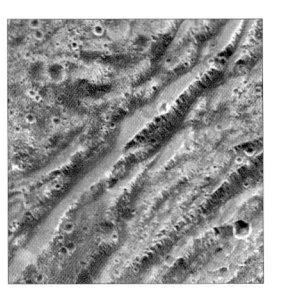

006 **Valentine present.** On Valentine's Day 2000, the NEAR-Shoemaker spacecraft entered the oblong asteroid Eros, named after the god of love. Eros measures 33 by 13 kilometres. The large impact crater is six kilometres across.
© *NASA/JPL/JHU-APL*

009 **Rocky vessel.** Phoebe is one of the outermost satellites of Saturn. Craters on its rocky surface, imaged by NASA's Cassini spacecraft, have been named after protagonists of the Greek story of the Argonauts. The huge crater at the upper right is called Jason.
© *NASA/JPL/SSI*

010 **Groovy Ganymede.** The largest planetary moon in the solar system, Ganymede is a world on its own, with mysterious tectonic grooves and small impact craters. The Jovian moon measures 5,262 kilometres across – larger than the planet Mercury.
© *NASA/JPL/Brown University*

013 **New Moon.** This mosaic of 18 images, captured in December 1992 by NASA's Galileo spacecraft en route to the giant planet Jupiter, shows our familiar Moon from a new angle: only the left part of this north polar view can be seen from the Earth.
© *NASA/JPL/USGS*

014 **Tick mark.** Looking like a gigantic tick, this bizarre volcanic structure on the surface of Venus measures some 66 kilometres across. Its west rim (left) appears to be breached by lava flows. This radar image was obtained by the Magellan spacecraft.
© *NASA/JPL*

016 **Lunar arena.** Captured in a dramatic oblique view by the unmanned space probe Lunar Orbiter 2 in the 1960s, the 93-kilometre wide impact crater Copernicus resembles a giant Roman arena. Copernicus is visible with the naked eye.
© *NASA/JPL/USGS*

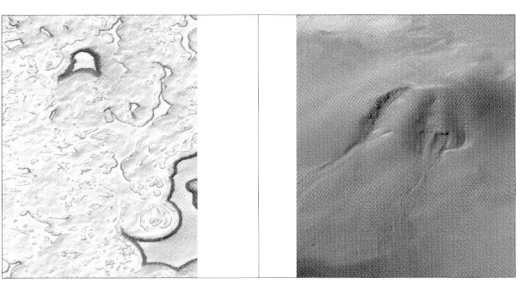

018 **Rainbow Bay.** The dark plain on the right of this photo is Sinus Iridum ('Rainbow Bay'), an old impact structure on the near side of the Moon. This photo was taken with the Wide Field Imager of the 2.2-metre telescope at the European La Silla Observatory.
© *ESO*

019 **Third dimension.** This perspective view of the giant impact crater Hale on Mars was created by the High Resolution Stereo Camera on board the European spacecraft Mars Express. The view is towards the northwest. Hale has a diameter of 136 kilometres.
© *ESA/DLR/G. Neukum (Freie Universität Berlin)*

020 **Twin peaks.** Two small hills rise above the Martian horizon in this panoramic image made by the Mars Pathfinder lander. The peaks are about 35 metres high. Pathfinder touched down in the rock-strewn area on Independence Day 1997.
© *NASA/JPL*

022 **Dry ice.** Closeup of the south polar cap of Mars during mid-summer, when all water ice has sublimated and only frozen carbon dioxide ('dry ice') remains. The dark rims of the raised plateaux are probably caused by dust that has been trapped in the ice.
© *NASA/JPL/MSSS*

023 **Dry run.** Gullies on the rim of a small impact crater on Mars suggest that water once flowed here. Scientists assume subsurface ice may occasionally seep through the crater wall, creating temporary flows that transport rocks and dust.
© *NASA/JPL/MSSS*

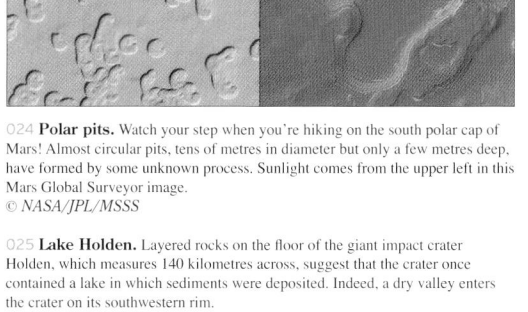

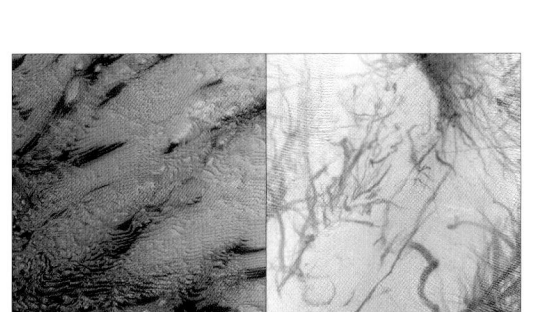

024 **Polar pits.** Watch your step when you're hiking on the south polar cap of Mars! Almost circular pits, tens of metres in diameter but only a few metres deep, have formed by some unknown process. Sunlight comes from the upper left in this Mars Global Surveyor image.
© *NASA/JPL/MSSS*

025 **Lake Holden.** Layered rocks on the floor of the giant impact crater Holden, which measures 140 kilometres across, suggest that the crater once contained a lake in which sediments were deposited. Indeed, a dry valley enters the crater on its southwestern rim.
© *NASA/JPL/MSSS*

026 **Newspaper rock.** Like pages of a book or a newspaper, hundreds of thin rocky layers are exposed at the bottom of an impact crater in Arabia Terra on Mars. The message they contain is that some form of sedimentation, either by water or dust, must have occurred in the past.
© *NASA/JPL/MSSS*

027 **Tornado alley.** On some spring and summer afternoons on Mars, little tornadoes sweep up fine dust as they move across the surface. Dozens of these so-called dust devils have left their trails on the floor of the Argyre basin, in the form of dark, narrow streaks.
© *NASA/JPL/MSSS*

028 **Direct hit.** A remarkable group of small impact craters in Arabia Terra on Mars may have formed when a meteorite broke up in the planet's atmosphere before slamming into the surface. The largest craters are about one kilometre wide.
© *NASA/JPL/MSSS*

029 **Crater rim.** The European spacecraft Mars Express captured the eastern rim of the huge impact crater Huygens. Named after Dutch physicist and astronomer Christiaan Huygens, the crater shows evidence of water run-off in the past.
© *ESA/DLR/ G. Neukum (Freie Universität Berlin)*

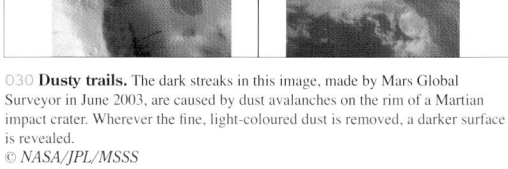

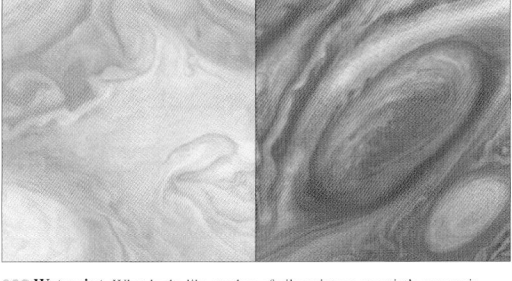

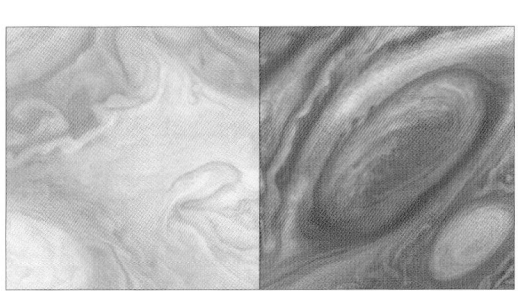

030 **Dusty trails.** The dark streaks in this image, made by Mars Global Surveyor in June 2003, are caused by dust avalanches on the rim of a Martian impact crater. Wherever the fine, light-coloured dust is removed, a darker surface is revealed.
© *NASA/JPL/MSSS*

031 **Night vision.** Using its thermal camera, NASA's space probe 2001 Mars Odyssey captured this enhanced false-colour scene at night. The image shows flow patterns near the edge of the northern polar cap, during summertime.
© *NASA/JPL/ASU*

032 **Wet paint.** What looks like strokes of oily paint on an artist's canvas is actually a natural-colour closeup of turbulent clouds in the atmosphere of Jupiter, just to the southeast of the Great Red Spot, which is by far the largest cyclone system on the planet.
© *NASA/JPL*

033 **Red Spot.** Discovered over three hundred years ago, Jupiter's Great Red Spot is a giant storm system large enough to contain two Earths side by side. The colours in this photo by the Voyager 2 spacecraft are slightly enhanced to bring out more detail.
© *NASA/JPL*

034 **Above Pathfinder.** Panoramic images made by NASA's Mars Pathfinder lander were processed to create this top view of the probe's landing site as it would appear from above. The black circle represents parts of Pathfinder not imaged by its cameras.
© *NASA/JPL*

035 **Grand canyon.** Valles Marineris on Mars is the largest canyon in the solar system. It is about 4,000 kilometres long, up to 200 kilometres wide and in some places over 6 kilometres deep. This mosaic of the canyon was shot by NASA's Viking orbiter.
© *NASA/JPL*

048 Radio power. In the core of the galaxy M87, a supermassive black hole creates a narrow jet of charged particles that emit radio waves. The giant blobs in this false-colour image are produced by the interaction of the jet with the surrounding gas.
© NRAO/NSF

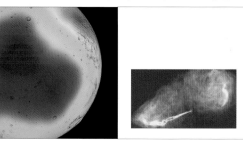

049 Subsurface ice. Looking down at the north pole of Mars, NASA's space probe 2001 Mars Odyssey detected subatomic particles called neutrons (colour-coded in blue) that are probably emitted by subsurface ice. Red and orange areas are relatively ice-free.
© NASA/JPL/GSFC/IKI

050 Blowing bubbles. The fierce stellar wind of a hot star (to the lower right of this picture) sweeps up interstellar material in a giant crescent-shaped bubble. The blue nebulosity is a false-colour representation of X-rays emitted by shocked and superheated gases.
© NASA/University of Illinois/San Diego State University/Mount Laguna Observatory/Y. Chu & R. Gruendl et al.

051 Evil axis. Hidden in the dusty yellow cloud at the centre of this image is a protostar, which reveals itself by two narrow, geyser-like jets, blown in the two and eight o'clock directions and producing greenish bow shocks along the way. Because of their catalogue number, HH666, the jets are known as the Axis of Evil.
© Nathan Smith and John Bally (University of Colorado) and NOAO/AURA/NSF

052 UV block. Thanks to the Earth's ozone layer, extreme ultraviolet radiation from the Sun, seen on this false-colour image by the SOHO spacecraft, is unable to reach our planet's surface, where it would destroy living cells. The image was made in May 1998.
© NASA/ESA

053 Pink necklace. Like a pearl necklace, a glowing ring of gas surrounds the remains of a star in the Large Magellanic Cloud that exploded in February, 1987. The gaseous ring has been ejected by the star in an earlier stage and is now energised by the radiation from the explosion.
© NASA/ESA/P. Challis and R. Kirshner (Harvard-Smithsonian Center for Astrophysics)/B. Sugerman (STScI)

042 Whirling clouds. Numerous anti-cyclonic cloud systems are visible in the atmosphere above Saturn's northern hemisphere. The largest are about 250 kilometres across. This false-colour image was made by NASA's Voyager 2 spacecraft in 1980.
© NASA/JPL

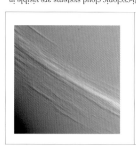

043 Blue planet. High-altitude clouds slowly drift across the methane-rich lower atmosphere of Neptune in this uncanny Earth-like view captured by Voyager 2 in August 1989. Neptune is the most distant planet ever visited by any spacecraft.
© NASA/JPL

044 Pseudo Saturn. Because of its appearance in small telescopes, this planetary nebula, catalogued as NGC 7009, is also known as the Saturn Nebula. The red blobs (the end points of the 'rings') are regions of low-density gas, ejected by the central star.
© Bruce Balick and Jason Alexander (University of Washington), Arsen Hajian (US Naval Observatory), Yervant Terzian (Cornell University), Mario Perinotto (University of Florence, Italy), Patrizio Patriarchi (Arcetri Observatory, Italy), NASA/ESA

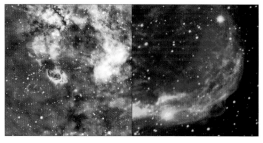

045 Venusian hills. Using radar data from the Magellan spacecraft, scientists constructed this elevation map of the surface of Venus, which is permanently veiled by clouds in its thick atmosphere. The equatorial highland is known as Aphrodite Terra, after the Greek name for Venus.
© NASA/JPL/MIT/USGS

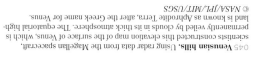

046 First light. Blue and yellow spots on this image represent tiny temperature variations in the cosmic background radiation, the oldest signal ever observed in the universe. This map by NASA's Wilkinson Microwave Anisotropy Probe is in a sense a picture of the Big Bang.
© NASA/WMAP Science Team

047 Blue bubble. Once upon a time, the star in the centre of this cosmic soap bubble (known as Abell 39) used to shine as bright as the Sun. Then it blew away its outer layers and contracted into a tiny white dwarf, which now energises the expanding nebula.
© WIYN/NOAO/NSF

036 Speedy moon. The Jovian moon Io sails above the cloud deck of its mother planet in this remarkably detailed Hubble Space Telescope picture. Io is slightly smaller than our own Moon. It orbits Jupiter every 1.8 days. The black dot at the left is Io's shadow.
© J. Spencer (Lowell Observatory), NASA/ESA

037 Waltzing Dione. Saturn's icy moon Dione floats high above the cloud deck. This picture was taken by the Cassini spacecraft, named after the French astronomer Jean-Dominique Cassini who discovered Dione in 1684.
© NASA/JPL/SSI

038 Parting shot. After it encountered Uranus in January 1986, NASA's Voyager 2 spacecraft took this dramatic picture of the backlit, crescent planet, with its distinctive blue-green hue, caused by methane in its atmosphere.
© NASA/JPL

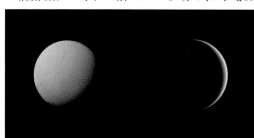

039 Bright moon. Saturn's icy moon Enceladus reflects almost 90 per cent of incident sunlight. It is actually the most reflective object in the solar system. Cryovolcanism has produced sinuous flows of frozen water, while meteoric impacts have punctured the surface.
© NASA/JPL

040 African enchantment. Delicate streamers and majestic loops of hot gas are caught in this composite image of the 2001 total solar eclipse in Zambia. Twenty-two separate images have been combined and processed to closely resemble the naked-eye appearance of the corona.
© F. Espenak, www.MrEclipse.com

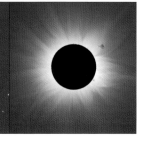

054 Icy freckles. Dark spots, each some ten kilometres across, pepper the icy surface of the Jovian moon Europa. Like the elongated cracks, these freckles and pits are probably filled with icy slush moving upwards. A liquid ocean is concealed beneath the frozen crust.
© NASA/JPL/University of Arizona/University of Colorado

055 Lava rules. Tvashtar Catena is a chain of active volcanoes on Jupiter's moon Io. In this false-colour image, a new lava flow glows orange-red because of its infrared (heat) emission. Older lava flows are black. Io's surface is coloured yellow by sulphur deposits.
© NASA/JPL

056 Colourful Culann. Red and green volcanic ash deposits and colourful lava flows make up this Galileo image of Culann Patera, an active sulphur volcano on Jupiter's moon Io. Colour variations have been enhanced. The smallest detail in the photo are 200 metres across.
© NASA/JPL

057 Pizza moon. Io, the innermost of the large moons of Jupiter, is the most volcanically active world in the solar system. Because of the numerous burn marks and lava flows on its surface, Io is effectively known as the pizza moon.
© NASA/JPL

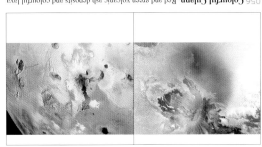

058 Giant flare. One of the largest solar flares in the past century occurred in early November 2003. The ultraviolet emission of the flare – the white area in this false-colour image – completely saturated the detectors of the US/European SOHO satellite.
© NASA/ESA

059 Golden rivers. In the hot, tenuous layers above a sunspot the size of the Earth, the Dutch Open Telescope at La Palma has captured magnetic flows of incandescent hydrogen gas, artificially coloured in this single-wavelength image.
© University of Utrecht/Rob Rutten

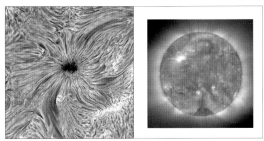

060 Solar handle. A huge prominence of magnetised gas shaped like a giant handle erupts from the sizzling surface of the Sun in this ultraviolet image by the US/European spacecraft SOHO. White areas on the Sun are hot active regions.
© NASA/ESA

061 Towering inferno. A one-million-degree hot fountain of hydrogen gas, higher than the diameter of the Earth, leaps up from the solar surface, supported by powerful magnetic fields. The image was taken by the American TRACE spacecraft.
© Stanford-Lockheed Institute for Space Research/Karel Schrijver

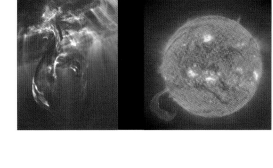

062 Nine rings. Narrow rings of ice and dust encircle the planet Uranus. This enhanced-colour view, captured by Voyager 2 in 1986, reveals subtle differences in composition between the various rings, nine of which have been labelled with numbers and Greek letters.
© NASA/JPL

063 Icy colours. Measurements by the Cassini spacecraft of ultraviolet sunlight reflected by Saturn's icy rings have been worked into this false-colour image. Red corresponds to relatively dirty ice; green colours indicate cleaner ice.
© NASA/JPL/University of Colorado

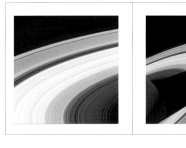

064 Blue crepe. Sometimes called the Crepe Ring, the low-density innermost region of Saturn's ring system (the C ring) has a different composition from the bright B ring, as evidenced by the blue hues in this false-colour photograph by the Voyager 2 spacecraft.
© NASA/JPL

065 Tidal waves. Wave-like structures in a thin sheet of very fine dust particles are created by tidal forces of small satellites in Saturn's ring system. The bright ribbon in the upper right is the narrow F ring, which is also shepherded by two small moons.
© NASA/JPL/SSI

066 Black oval. Saturn, the second-largest planet in our solar system, casts a pitch black shadow on its delicate ring system in this photo by the Cassini spacecraft. Just outside the main ring system is the narrow F ring, discovered in 1979 by Pioneer 11.
© NASA/JPL/SSI

067 Running rings. Just before it entered orbit, NASA's Cassini spacecraft captured this striking natural-colour image of Saturn's ring system. The countless narrow ringlets and gaps are produced by tidal forces from small satellites.
© NASA/JPL/SSI

068 Edge-on Saturn. This Hubble Space Telescope photograph of Saturn, made in 1996, shows the ring system of the giant planet almost edge-on. Cloud bands in the planet's atmosphere are created by its fast rotation of just over 10.5 hours.
© NASA/Hubble Heritage Team (STScI/AURA)/R.G. French (Wellesley College), J. Cuzzi (NASA/Ames), L. Dones (SwRI), and J. Lissauer (NASA/Ames)

070 Majestic beauty. Thick dust lanes in the central plane encircle the radiant central parts of this majestic galaxy. Known as M104 or the Sombrero galaxy, it contains 800 billion stars and is located at some 28 million light-years from Earth.
© NASA/ESA/Hubble Heritage Team (STScI/AURA)

073 **Grand spiral.** The 8.1-metre Gemini North Telescope at Mauna Kea, Hawaii, captured this image of the spiral galaxy M74 in the constellation Pisces the Fishes. The galaxy resembles our own Milky Way. It is located at a distance of some 35 million light-years.
© *Gemini Observatory/GMOS Commissioning Team*

072 **Back spin.** Spiral galaxies usually rotate in the direction that would wind up the spiral arms. However, measurements of this galaxy, NGC 4622, suggest it is spinning backwards, probably due to a merger with another galaxy in the past.
© *NASA/ESA/Hubble Heritage Team (STScI/AURA)*

075 **Radiant ring.** Bright star-forming regions form a luminous ring surrounding the core of this serene spiral galaxy at a distance of 45 million light-years in the southern constellation Fornax the Oven. The galaxy core probably conceals a supermassive black hole.
© *ESO*

074 **Barred spiral.** The spiral arms of this stately galaxy, catalogued as NGC 1300, emanate from the end points of an elongated bar of stars in the centre. This mosaic consists of several Hubble Space Telescope photographs.
© *NASA/ESA/Hubble Heritage Team (STScI/AURA)*

077 **Chance alignment.** By pure coincidence, these two galaxies happen to be perfectly aligned with Earth. As a result, astronomers are able to actually see the dark material in the smaller foreground galaxy, silhouetted against the bright backdrop of the other.
© *NASA/ESA/Hubble Heritage Team (STScI/AURA)*

076 **Evil eye.** Created by the merging of two spirals, this Yin Yang-like galaxy, known as M64, sports a broad dark band of obscuring dust, as well as a sparkling of young star clusters, formed as a result of shock waves in the interstellar gas.
© *NASA/ESA/Hubble Heritage Team (AURA/STScI)*

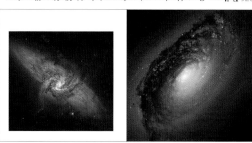

078 **Telling tails.** Comet Hale-Bopp, which graced our skies in the spring of 1997, showed two distinct tails: a yellow-white dust tail and a narrow blue gas tail. Both tails form under the influence of sunlight and solar wind.
© *Akira Fujii/DMI*

079 **Neat comet.** In the spring of 2004, comet NEAT graced the northern hemisphere skies. The comet is named after the Near-Earth Asteroid Tracking program with which it was discovered. Like other comets, it is a small chunk of evaporating ice and dust.
© *T. A. Rector (University of Alaska Anchorage), Z. Levay and L. Frattare (Space Telescope Science Institute) and WIYN/NOAO/AURA/NSF*

080 **Tadpole race.** Dense knots of gas are eroded by the energetic radiation of a star, leaving faint faint tails reminiscent of comets in our solar system. These cosmic tadpoles populate the outer regions of the Helix Nebula in the constellation Aquarius the Water Bearer.
© *Robert O'Dell and Kerry P. Handron (Rice University, Houston, Texas)/NASA/ESA*

081 **Cosmic blizzard.** Some 8,000 hot, blue stars swarm around the core of the small elliptical galaxy M32, a satellite of the Andromeda galaxy, like snowflakes in a blizzard. The velocities of the stars reveal the presence of a black hole at the galaxy's core.
© *NASA/ESA/Thomas M. Brown, Charles W. Bowers, Randy A. Kimble and Allen V. Sweigart (GSFC/Henry C. Ferguson (STScI)*

082 **Hail storm.** Atoms and molecules have frozen out and condensed into hail stones in the inner regions of the Bug Nebula, ejected by a superhot dying star, hidden behind dust clouds in the upper right corner of this image by the Hubble Space Telescope.
© *ESA/NASA/Albert Zijlstra*

083 **Nested wheels.** At 2,000 light-years from Earth, the Hubble Space Telescope captured this striking image of a planetary nebula known as the Spirograph Nebula for obvious reasons. No one knows how to explain the subtle texture of the slowly expanding gas cloud.
© *NASA/ESA/Hubble Heritage Team (STScI/AURA)*

084 **Supernova shock.** The thin green ring in this X-ray image marks the location of a shock wave, produced by a supernova explosion in the late 17th century. This composite false-colour image of the supernova remnant, Cassiopeia A, also shows an energetic jet.
© *NASA/CXC/GSFC/U. Hwang et al.*

085 **Mortal remains.** The expanding shell of Kepler's supernova, which exploded in 1604, is seen here in infrared radiation (red), visible light (yellow) and low- and high-energy X-rays (green and blue, respectively).
© *NASA/ESA/JHU/R. Sankrit and W. Blair*

086 **Flower power.** A huge but extremely faint halo of gas, shaped like a flower, surrounds the well-known Cat's Eye Nebula. The halo measures three light-years across. It was captured by the Nordic Optical Telescope at La Palma.
© *Nordic Optical Telescope and Romano Corradi (Isaac Newton Group of Telescopes, Spain)*

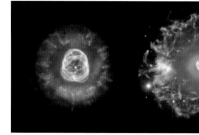

087 **Face value.** Vaguely resembling a human face surrounded by a fur parka, this planetary nebula is known as the Eskimo Nebula. This Hubble image was the first one taken by the space telescope after a successful servicing mission in December 1999.
© *NASA/ESA/Andrew Fruchter (STScI)/ERO team (STScI/ST-ECF)*

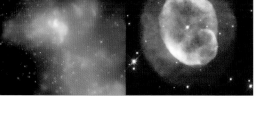

088 **Colour wizardry.** Using four black-and-white Hubble photos, made at different wavelengths, astronomers created this false-colour view of a planetary nebula known as NGC 5979. It shows the final stage in the life of a sun-like star.
© *ESA/ESO/NASA*

089 **X-ray vision.** Clouds of superheated gas in the centre of the Milky Way, with temperatures as high as ten million degrees, are visible in this false-colour X-ray image. The gas is probably heated by shock waves from supernova explosions.
© *NASA/CXC/UCLA/MIT/M. Muno et al.*

090 **Exploding galaxy.** Some ten million years ago, a giant explosion in the galaxy Centaurus A produced energetic particle jets and giant arcs of hot, X-ray emitting gas. This image combines observations at radio wavelengths (pink and green), visible light (orange and yellow) and X-rays (blue).
© NASA/CXC/M. Karovska et al./NRAO/J. Van Gorkom et al./J. Condon et al./Digitized Sky Survey/UK Schmidt Telescope/STScI

091 **Purple rain.** Hot hydrogen gas, colour-coded purple in this composite image, rains up and down from the core of the galaxy M82, which is ablaze with star formations, probably as a result of a close encounter with another galaxy in the recent past.
© Mark Westmoquette (University College London), Jay Gallagher (University of Wisconsin-Madison), Linda Smith (University College London), WIYN/NSF, NASA/ESA

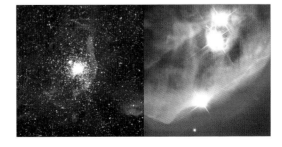

092 **Transparent cone.** Using its infrared camera NICMOS, the Hubble Space Telescope was able to peer through the absorbing dust clouds at the tip of the so-called Cone Nebula, revealing newly born stars invisible to the human eye.
© NASA/ESA/NICMOS Group (STScI, ESA) and the NICMOS Science Team (University of Arizona)

093 **Double cluster.** Flanked by a red veil of nebulosity in the Large Magellanic Cloud is a remarkable double cluster of stars. The smaller cluster may have formed as a result of shock waves from supernova explosions in the large cluster.
© ESO

094 **Magellanic cradle.** Known as N11, this pink cloud of glowing hydrogen gas is a massive cradle of new stars, located in the Large Magellanic Cloud, a small satellite galaxy of the Milky Way at 169,000 light-years distance. Many newly-born stars are also visible.
© C. Aguilera, C. Smith and S. Points/NOAO/AURA/NSF

095 **Light echo.** The titanic explosion of a remote star at a distance of 20,000 light-years reverberates through the cosmos in the form of eerie light echoes: light from the explosion is reflected by clouds of gas and dust that have been ejected by the star during earlier explosions.
© NASA/ESA/Hubble Heritage Team (AURA/STScI)/ESA

096 **Tarantula's labour.** Spawning millions of new-born suns, the giant Tarantula Nebula in the Large Magellanic Cloud would be much brighter than the full Moon if it was moved to the location of the Orion Nebula. This mosaic consists of various Hubble Space Telescope images.
© ESA/NASA/ESO/Danny LaCrue

097 **Fiery aftermath.** Shaped by shock waves and magnetic field, filaments and tendrils of hot gas mark the spot of a catastrophic explosion in the year 1054. The lower one of the two relatively bright stars is the pulsar that was produced during the supernova.
© NASA/ESA/Hubble Heritage Team (STScI/AURA)

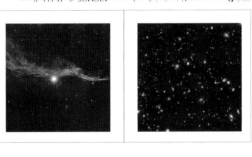

098 **Deep space.** Using its infrared camera NICMOS, the Hubble Space Telescope captured the faint glow of hundreds of extremely remote galaxies in the so-called Hubble Ultra Deep Field, where astronomers look back in time to the early history of the universe.
© NASA/ESA/R. Thompson (University of Arizona)

099 **Thin Veil.** Tenuous wisps of gas are all that is left of a catastrophic supernova explosion that happened thousands of years ago in the constellation Cygnus the Swan. Known as the Veil Nebula, it passes behind an unrelated foreground star.
© T. A. Rector/University of Alaska Anchorage and WIYN/NOAO/AURA/NSF

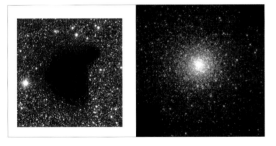

100 **Stellar beehive.** Located at some 28,000 light-years from Earth, M80 is one of the densest globular clusters known. It contains many hundreds of thousands of stars, most of which are at least twelve billion years old.
© NASA/ESA/Hubble Heritage Team (AURA/STScI)

101 **Dark cloud.** At just 410 light-years distance, Barnard 68 is one of the nearest dark clouds in the Milky Way. The dust cloud, from which new stars will form in the future, completely obscures the light of stars in the background.
© ESO

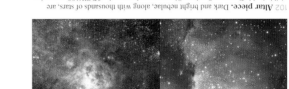

102 **Altar piece.** Dark and bright nebulae, along with thousands of stars, are visible in this European Southern Observatory photograph of RCW108, a massive star-forming region in the southern constellation Ara the Altar.
© ESO

103 **Distant spider.** More than ta thousand light-years across, the Tarantula Nebula in the Large Magellanic Cloud (so named because of its shape) is one of the largest star-forming regions known. Despite its distance of 170,000 light-years, it is easily visible with the naked eye.
© ESO

104 **Stellar sphere.** 47 Tucanae is one of the largest and brightest globular star clusters in the sky, in the southern constellation Tucana, at a distance of over 13,000 light-years. This image was made as part of the near-infrared 2MASS survey.
© 2MASS/T. Jarrett

105 **Twinkle twinkle.** Tiny white dwarf stars pepper the inner regions of M4, a globular cluster of stars in our own Milky Way galaxy. The stars are almost as old as the universe: 12 to 13 billion years. The photo only shows a tiny region of the cluster.
© NASA/ESA/H. Richer

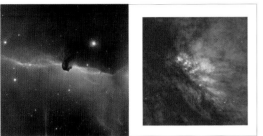

106 **Baby boom.** A giant burst of star formation in the inner 1,000 light-years of the galaxy NGC 253 creates violent shock waves in the surrounding gas and dust. The new baby stars are obscured by clouds, but the starburst leaves its mark in the immediate vicinity.
© Carnegie Institution of Washington

107 **Hunter's horse.** Located in the constellation Orion the Hunter, the dark protrusion in this wide-angle photographs is known as the Horsehead Nebula. It is part of a large cloud of dust, which blots out the delicate glow of hot wisps of gas in the background.
© T. A. Rector (NOAO/AURA/NSF) and Hubble Heritage Team (STScI/AURA/NASA)

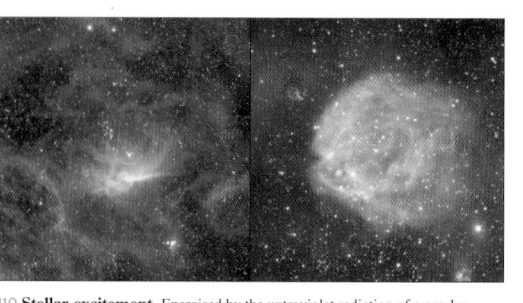

108 Hatching stars. Embedded in blue nebulosity, large numbers of small and light-weight embryonic stars still await ignition of their nuclear furnaces. They were discovered by the Hubble Space Telescope in the Small Magellanic Cloud, a satellite of our Milky Way galaxy.
© *NASA/ESA/A. Nota (STScI)*

109 Stellar nursery. Tens of thousands of stars were born over the past two million years in this giant star-forming region in the Orion Nebula. The four bright, massive stars at the top of this near-infrared false-colour image constitute the Trapezium cluster.
© *ESO/Mark McCaughrean*

110 Stellar excitement. Energised by the untraviolet radiation of a nearby extremely hot star, these gaseous filaments in the Large Magellanic Cloud (a satellite of our Milky Way galaxy) glow at the characteristic wavelengths of hydrogen, helium and oxygen.
© *ESO*

111 Hottest star. AB7, the binary star in the centre of this colourful bubble in the Small Magellanic Cloud, is one of the hottest stars known, with a surface temperature of a staggering 120,000 degrees. Its fierce stellar wind blows surrounding material away.
© *ESO*

112 Cosmic fireworks. A salvo of supernova explosions in the galaxy NGC 1569, starting some 25 million years ago, carved huge bubbles in the interstellar material, while shock waves led to the formation of new clusters of bright, young stars.
© *ESA/NASA/Peter Anders (Göttingen University Galaxy Evolution Group, Germany)*

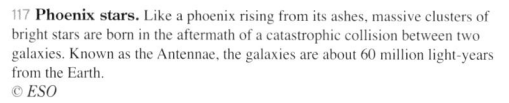

114 Core business. A supermassive black hole at the core of our Milky Way galaxy lights up its surroundings in X-rays, captured in this false-colour image by NASA's Chandra X-ray Observatory. Many flaring X-ray stars are also visible in this two-week exposure.
© *NASA/CXC/MIT/F. K. Baganoff et al.*

115 Bug feature. Known as the Bug Nebula because of its shape, NGC 6302 is a planetary nebula, ejected by a dying star like the Sun. This image was taken with the 3.6-metre telescope at the European Southern Observatory at La Silla, Chile.
© *ESA/NASA/Romano Corradi (Isaac Newton Group of Telescopes, Spain)*

116 Flaming star. AE Aurigae is the name of the blue star at the centre of this image. Its prodigious energy output heats up the surrounding clouds of gas and dust, causing them to glow at wavelengths characteristic of hydrogen, oxygen and sulphur atoms.
© *T. A. Rector and B. A. Wolpa/NOAO/AURA/NSF*

117 Phoenix stars. Like a phoenix rising from its ashes, massive clusters of bright stars are born in the aftermath of a catastrophic collision between two galaxies. Known as the Antennae, the galaxies are about 60 million light-years from the Earth.
© *ESO*

118 Blue ribbon. The Hubble Space Telescope captured this tiny portion of the Cygnus Loop, the tenuous remains of a supernova explosion that happened some 5,000 years ago. The date has been confirmed by measurements of the ribbon's motion.
© *ESA/Digitized Sky Survey (Caltech)*

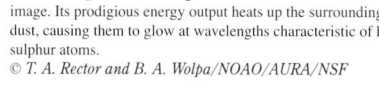

127 Stately Saturn. A staggering 126 images by NASA's Cassini spacecraft were assembled to create this impressive mosaic of the ringed giant planet Saturn. It is the most detailed image of Saturn and its rings available to date.
© *NASA/JPL/SSI*

128 Lasting impression. It will be millions of years before the footprints of Apollo 11 astronauts Neil Armstrong and Buzz Aldrin are eroded away, since there is no wind or water on the Moon. Mankind has truly made a lasting impression on the cosmos.
© *NASA*